THE FISH DOCTOR:
AUTOBIOGRAPHY OF A
WORLD FISH PARASITOLOGIST

Glenn L. Hoffman, PhD

The Fish Doctor

Autobiography of a World Fish Parasitologist

Glenn L. Hoffman, PhD

Author of *Parasites of North American Freshwater Fishes*,
published in 1999 by Cornell University Press

Zea E-Books
Lincoln, Nebraska
2011

Cover illustration: Trout fry afflicted with *Myxobolus cerebralis* in a detail from a photograph by the author.

ISBN 978-1-60962-007-3   e-book
ISBN 978-1-60962-006-6   paperback

Copyright © 2011 Estate of Glenn L. Hoffman.

Set in Goudy Old Style types.

 Zea E-Books are published by the University of Nebraska–Lincoln Libraries.

Contents

Editor's Note		vii
1.	Ancestors and Parents	1
2.	Childhood	8
3.	College Days	14
4.	University of Iowa, 1939-1942	17
5.	College Friends and Roommates	22
6.	My Brother, Melvin	26
7.	World War II	29
8.	University of Iowa, PhD Program, 1946-1950	47
9.	Carolyn Elise Wilson Hoffman	54
10.	Carolyn's Family and Ancestors	59
11.	University of North Dakota, 1950-1957	65
12.	Fish Parasitology Researcher, USFWS, Leetown, WV	72
13.	Scientific Collaboration, 1960-1975	86
14.	The Hoffman Offspring	88
15.	U.S. Fish Culture Research Station, Stuttgart, AR	91
16.	Those Who Helped Me	96
17.	Foreign Trips to Conferences and Visitations to Fish Parasite Labs	100
18.	Domestic Vacation Trips, Some Coupled to Parasitology Meetings	125
19.	Retirement, 1985-[2010]	133
20.	Epilogue	141

Appendix A
   Lifetime Publications of Glenn L. Hoffman, Sr.
      as Senior Author   143

Appendix B
   Lifetime Publications of Glenn L. Hoffman, Sr.
      as Junior Author   147

Editor's Note

The author, my father, completed a rough draft of this manuscript before his death in February, 2010. He had spoken to me about his plans to incorporate photographs, but he did not leave any indication of which photographs (out of the thousands in his archives) he considered appropriate. It has fallen to me to select those photographs which illustrate the people and places he has mentioned in the document, and to revise the text. My choices are inevitably different from those my father would have made. As much as possible I have retained the language he used in the text, and have resisted the temptation to add anecdotes he told me over the years, or my own recollections of events and people. All figure captions, however, were added by me.

I would like to express sincere gratitude to Dr. Scott Gardner, Curator of Parasitology at the University of Nebraska State Museum, for agreeing to publish this manuscript electronically, and to Gail Littrell for word-processing the rough draft.

G. Lyle Hoffman
June, 2010

1. Ancestors and Parents

My paternal grandparents, Ernest Hoffmann and Mary Puff of Silesia, Germany (now Poland), were married in Dubuque, IA, in 1890. They raised a family in Hazelton, IA — Mayme Hoffman Schilling, Hilda Hoffman Oetker, Ernest (my father), Frank, Alma Hoffman Destival, Leo, Leona Hoffman Wilson, Gertrude Hoffman Linvolle, and Lillian who never married. In 1980 Jeanette Hartman Cannon, a relative, self-published the genealogy up to that time of the Joseph and Johanna Puff family which includes my paternal grandparents and my parents Ernest James Hoffman and Viola Hall Hoffman. This publication is available only by photocopy. My mother, Viola Mildred Hall Hoffman, was the daughter of Albert N. Hall and Pauline Davis Hall who also had a son, Gerald (Bunny) Hall, who moved to California and married Doris (maiden name unavailable). Albert N. Hall then married Mary Anna Catherine Schurman. Their children were (are) Voretta (deceased), born 1909; Esther, born 1912; Burdette (deceased), born 1915; Thelma, born 1917; Julia, born 1912; Robert (Toby) and Golda, all half-sisters or half-brothers of my mother, Viola. Ernie, as my father was usually called, and Viola, my mother, worked for a substantial farmer named Glen Jakway near Aurora, IA, about eight miles east of Dad's birthplace, Hazelton, IA. They were married in 1915 and rented a nearby farm where my brother, Melvin, and I were born, he in 1917 and I in 1918. There was a nice spring near the house where apparently native Americans had camped long ago. My dad found many Indian arrowheads there. The farm was very near the headwaters of the Buffalo River (Creek) and I have a faint recollection of being there when my dad was fishing. Our dog chased a rabbit and my brother shouted "He stole my dog." So maybe my love of streams and fishing began there at a very early age! My family and I visited my birth site about 1980. The farm had become a county park and the house that Melvin and I had been born in had become the park headquarters—so our birth house lives on. Many years later I caught some nice small bass on the same stream somewhat downstream from my birth place.

Figure 1. My birthplace as it appeared in 1968. Photo by the author.

My Parents, Ernie and Viola

My father and mother were hard working Iowa farm folks before the age of large farm machines and household conveniences such as electric washing machines and dish washers. The farms they rented and the one they bought about 1940 were small by comparison to those of today. Most of them were about 120 acres, suitable to farming with four or five horses. Dad had learned well from his employer, Glen Jakway, about 1917, and he learned well from publications on farming methods, such as rotation of crops, value of legumes (soy beans and alfalfa) and composting all left over plant materials and waste from the barns. His farm was always well diversified with crops including hay for the cattle and horses, corn, oats, soy beans, timothy, alfalfa, and clover. There were no cash plant crops; all was feed for the livestock—cattle, pigs, horses, chickens, and once two goats. The cattle were mostly for milk, sometimes for market; the pigs for market; horses for work, chickens for eggs and capons for market. Dad was a normal-sized man, but very strong. At one point, about 1934 or 1935, in addition to all the farm work, he worked part time for the local highway department, im-

Figure 2. My parents in 1943. Photo by the author.

proving the nearby road connecting Lamont and the Backbone State Park very near our home.

My brother, Melvin, and I helped with the hand-milking before breakfast and after supper. Dad usually got up about 4:00 am, and we struggled up a little later. When we came home from school, my brother and I helped Dad. There was little time for school homework.

My mother, Viola Mildred, was the daughter of Albert and Pauline Hall. They also had a son, Gerald, who spent most of his life in California. Albert and Pauline were divorced and remarried, he to Anna (maiden name unknown to me) and she to a man named Wakeman; both lived in Des Moines, IA. I don't know their genealogy, but I recall my mother saying that Albert's ancestors were from England. I recall an incident at my maternal Grandmother's place when I was very young. On a walk near their house I met two boys who had been fishing in the Des Moines River and had two medium-sized carp. I bought the fish for a dime and gave them to Grandmother Pauline who cooked them.

My mother did all the house work including clothes washing in a hand-powered tub and house cleaning entirely by hand. She cared for

Figure 3. Ernie Hoffman plowing in 1939. Photo by the author.

the chickens—feeding, cleaning, and collecting the eggs. In summer she raised a large vegetable garden, and still found time for flowers. She played the piano "by ear."

Also remarkable was her ability to prepare noon meals for the oats threshers when they came to our farm. Before field combines became popular, the farmers in certain areas would band together to hire a threshing machine that would go from farm to farm. The neighbors, twenty or more men, would be the threshing crews. My mother, usually alone, would prepare the huge dinner meal for them when they came to our farm—a lot of work for a small woman.

I think my parents had normal tempers, and they upheld work habits with good composure. I do recall three incidents that irritated my dad greatly: Once a neighbor farmer and he walked to the Lamont Creek, about a half mile away, and snared spring-run suckers with wire snares at the end of bamboo poles. They caught about 15 or 20; most were a nice size, but some were smallish. Before parting, the neighbor divided the fish—one for me, one for you, etc., but somehow Dad ended up with all the smaller ones. Dad was not happy. About the same time, the same man went trout fishing in another stream, about two miles away, and caught his legal limit of trout because the stream had been recently stocked with proper-sized trout. He returned home, deposited his catch, then returned to the stream and caught a second limit. My dad was not happy with that one!

Figure 4. The farm my parents bought. Photo by the author, dated 1943.

Dad's brother, Frank, was in WWI overseas. When he returned, about 1920, he needed money. Dad loaned him $100 (year 2006 equivalent about $2000, I think). About 1935 my uncle's finances finally became substantial, so Dad asked for the return of his $100. The uncle refused, saying it was past the "statutes of limitation." Not fair, I think. My father was very angry. The other thing I recall about Uncle Frank was that during WWI, while he was stationed in France, his outfit had belly-crawling exercises in a field that had been spread with livestock manure.

My parents were farm renters (1917-1939) until they bought a farm near Lamont, IA. The first rented farm was about three miles south of Aurora, near Buffalo Creek, and belonged to Glen Jakway, Dad's farm boss. The second was the Hamlet Farm about two miles north of Lamont. The third was about six miles north of Lamont on the same road. The fourth was the Millet, now Hebrank farm, about one mile west of the West Gate of the Backbone Park, about four miles east of Lamont and situated between the Lamont Creek and the upper Maquoketa River. I mention these streams because eventually I could walk to either stream, catching smallmouth bass in the Lamont Creek and

Figure 5. Folk art carved by Ernie Hoffman, some finished and some not. These examples are in the possession of my daughter-in-law, Kathy Zadorozny. Photo by Lyle Hoffman.

trout in the Maquoketa. For my seventh and eighth grades I walked a mile north to a country one-room school, and later my brother and I walked about a mile south to the high school bus stop.

During this period a close-by neighbor had a new barn built. The sides and ends were made separately, and when they were ready neighbors came to help raise them. This was called a barn raising, a neighborly event.

About 1974 my parents sold their farm to a neighbor, Harry Stricher, who took good care of it which pleased them very much. They bought a house with three acres on the edge of Strawberry Point, northeast Iowa. Mother still had housework, of course, but she also had time for nice flower gardens again.

Dad had a large vegetable garden and was able to sell some surplus to others. I recall seeing him doing his gardening sitting on a stool because of arthritis. He also did a lot of carving. He made excellent farm animals, doll furniture, etc., for the neighbor children. The latter helped save his life; he had a major stroke while in the basement

ANCESTORS AND PARENTS

and Penny Henke, the little neighbor girl, whom Dad made toys for, found him, ran home to tell her mother, Mary Ann Henke, who got him to the hospital on time. Ironically, that nice little girl died some time later of esophageal cancer. Dad had also carved a complete farm scene which he gave to the local Strawberry Point, IA, Museum. At last account, 2007, it was still there.

Dad died in 1984 at age 90. Mother died at age 91, a year later. Both are buried in the County Corners Cemetery about a mile west of the Backbone State Park, not far from the farm they used to own. Some of mother's ancestors, the Halls, are also buried there.

2. Childhood

Most of us remember a few events of very early childhood. I remember three: I recall my father pretending to try to catch me by chasing me around a large piece of furniture in our house. It was fun. Another time my brother, Melvin, one and a half years older, and I were in the back of a horse-drawn buggy, waiting for a trip to start. There was a carpenter's hammer on the floor. I picked it up and whopped my brother on his head; fortunately no great harm was done. Also at that early age I recall my girl cousins of similar age pressing flowers under their garters. I remember that Dad planted some popcorn for my brother and me to care for and to sell to the popcorn vendors who popped it and sold it on Saturday nights, the big shopping night for farmers.

In high school I excelled in wrestling and in trying to find time to go fishing. At the end of my senior year, the well-liked wrestling coach, Mr. Rulifson, organized a "round robin" tournament, with the winner of the 95 and 105 lb weights match advancing and so forth. I, at 125 lbs, advanced to finally beat the heavyweight so I became a champion. I had won fourth place in the area open tournament at Cedar Rapids, Iowa.

I also recall being in senior class play. The leading lady, my neighbor, Virginia Vanek, was taller than I, the leading man, so she had to bend a little for me to kiss her!

During high school days I was a bit of a cutup and did some crazy things such as "horsing" around during baseball practice and getting kicked off the team. The principal became so much annoyed by such antics that, although I was captain of the wrestling team, he disallowed my senior wrestling captain's letter. Driving home from school in our Model A Ford one evening, I saw a small field of shocked corn. That system was common then. No one was around so I drove into the field and drove down a row of shocked corn, smashing the shocks to the ground. It was not a wise thing to do. After WWII when I had a little money, I sent the farmer a check for my estimated damage. Like a good neighbor, he never cashed it. Fortunately I soon outgrew such crazy antics.

I enjoyed the outdoors very much and often went for long walks to the Lamont Creek and then up or down its banks. Once I was on a small cliff overhanging the stream and saw three nice carp swimming

CHILDHOOD 9

Figure 6. The author (right) and his brother Melvin in a studio portrait dated 1921. Photo by Schneider Studio of Oelwein, IA.

at the surface. I don't know what they were doing, but I had a 22 caliber rifle and shot all three in the head. It is unusual to shoot fish in the water because water usually deflects the trajectory of the bullet. Carp from a nice clear stream were good to eat. Carp is not very popular in the US, but is considered good eating in Europe.

Before I learned to flyfish, I learned how to snag suckers (fish) with a treble hook in the small stream near our home. Suckers, although bony, were good eating. At about age sixteen, I accidentally learned a fisheries biology lesson. In the spring I was on a cliff overlooking the Lamont Creek, a warm water stream in summer. There was a nice pool there where the water had cut under the cliff. I saw several nice trout there where they were not expected to be, but the water was still cool. With earthworms for bait I caught three or four of the trout. The trout had been planted in the Maquoketa trout stream several miles away. I have since learned that, depending on rainfall, the planted trout may move considerable distances from the planting site, upstream or down. In Iowa, stocked trout have even been caught in the "warm" Mississippi River in the spring.

Trapping fur animals was a large part of my early life. I recall the first mink my brother and I caught. It was in a small flowage in a small slough. A cold snap froze everything there and the mink was frozen in the ice. I recall another mink I caught when I was very young; I needed a new winter coat and the mink pelt brought enough to buy me a new sheepskin coat. Even to this day, about seventy years later, the shoulder that carried the weight of the heavy trapping bag sags about two inches lower than the other!

During my high school days I became very interested in trapping fur animals because I needed money and, early on, I recognized that it could be a challenge, particularly catching mink and fox. I eagerly read what little literature I could find on the subject. Catching wily mink was a real challenge. In small waterways and edges of small streams one could, if lucky, visualize the path the mink might take while hunting, and place the trap accordingly. However, mink traveled more often on land adjacent to the water so one ought to be able to catch them there also. But, in all the trapping I did, I caught only two on land. One was on a narrow ledge on a cliff beside the water. Because they usually lope along they could easily jump over the trap. The literature I read suggested putting a small limb across the path several inches high so the mink would have to jump or step over it. It worked at that place, and I did similarly to catch fox, both red and gray, especially where the path was through a break in a fence. Now, at my age, I would not want to harm the wild animals.

I enjoyed the trapping very much and had only two unhappy incidents. Once, someone stole a mink from my trap, and the other event was equally as bad. At one very promising spot under a little bridge over a very small stream where a mink was almost certain to go, I thought I should ask the nearby farmer for permission, even though the trap would be on highway property. I asked, and he said if I caught a mink there he should have half the value of the pelt. I agreed. A few

Figure 7. Backbone State Fish Hatchery in 1943. Photo by the author.

days later there was a muskrat, not a mink, in the trap. Another farmer passed by before I got there and told the other farmer that Glenn had a MINK in the trap. When I told the first farmer it was a muskrat, he accused me of lying! A little misinformation can do a lot of harm. When I caught a skunk, no matter how careful I was, my classmates knew it!

During my senior high school days I had dammed the small spring water flow on our farm, thinking I would stock it with fish. In the spring, when it was still cool, I caught a nice smallmouth bass in the Lamont Creek, ran home with it and "stocked" my impoundment. That fish lived there until a small flood carried it away, perhaps back to the Lamont Creek. That was my first fish culture attempt. Later I did the same with two small trout from the other stream, the trout stream. I don't recall what happened to them. That was perhaps the start of my fish culture, fish parasitology career.

Later during this period I worked summers at the nearby Backbone State Fish Hatchery as a helper, feeding trout, cleaning ponds, mowing grass, and stocking trout in Iowa streams. I recall the first time I was entrusted to do the stocking alone at a stream many miles away. My boss was Bob Cooper, a wonderful superintendent. I was able to flyfish for trout in some of those streams and also to flyfish for smallmouth bass in three other streams, some miles away.

Figure 8. Portrait of the author as a high school graduate in 1936. Photo by Dupont of Waterloo, IA.

The Backbone State Park Beaver Story

About 1937 the Iowa State Conservation Dept. decided that the large Backbone State Park near Strawberry Point should have beaver. There hadn't been beaver in that area for many years. The Dept. obtained a pair from a nearby state and placed them in a man-made beaver house on the stream that comes from Richmond Springs nearby.

The entire park was well wooded. The maneuver was "successful" because the beaver reproduced well. They continued to reproduce, doing well in the park, and then, some years later, migrating to other streams that were not so well wooded. In some places, they destroyed what few trees were there. I have no specific information available but, from what little information I have, they must have migrated to a wide area, doing much damage. The moral, live and learn!

After high school, Paul Young (deceased), former school superintendent of my school, Lamont High, who had taken a job at a larger school in Davenport, IA, invited me to join him and a group of high school boys from Davenport on a camping and fishing trip to Ely in northern Minnesota. Further, following that I could join his family in Davenport, start college at St. Ambrose College and help with the Davenport High School wrestling team. All of this was very kind of him and a great help to me. At the lake near Ely, I caught some large rock bass, but a huge northern pike severed my line because I hadn't learned to use wire leaders.

3. College Days

In Davenport, my roommate was Paul Young's nephew, Bill Young (deceased). He was a very good roommate. I didn't do very well at St. Ambrose College, but I have always been grateful for the excellent Music Appreciation course I took there. Being short of funds, I dropped out of St. Ambrose at the end of my freshman year, worked at the Backbone Fish Hatchery that summer and trapped fur animals again. It was during that period that my parents bought their farm near Lamont, IA.

I have always been very grateful to Paul Young for giving me the boost to start college. While living with the Youngs, I became interested in photography, learning how to compose pictures and to develop film and enlarge photos, all of this in the Youngs' house. My big thrill was winning second place in a local newspaper photo contest. My photo was that of a winding staircase up a tower.

I had no scholarships and so was to work my way through college. I had saved enough for my freshman tuition but, of course, needed money for essentials. Bill and I each gave eight dollars per week to the Young family for room and usually the main meal, so I had to work part-time. At various times I was an usher at a movie theater, a Sunday deliverer of hard rolls (bread), and the longest time as a clerk in a sporting goods store—the latter because of my interest in fishing. There I received the going wage of a part-time clerk—eight dollars per week. I recall one time when I had only thirty-five cents which was just enough for a hot beef sandwich.

The owner of the Sporting Goods store was Jim Creighton, a kind man. Once when I had good luck fishing in the nearby Mississippi River and no way to cook my catch, he had the fish cooked at a nearby restaurant at no cost to me. Also, after WWII, I needed a new fly rod which I ordered from him, and he mailed it to me at a nice discount.

During the year in Davenport while attending St. Ambrose College, I continued my interest in wrestling and became the captain of the YMCA wrestling team. We had four meets, winning two, losing one, and ending in a deadlock in the fourth. We entered the Waterloo, IA, Open tournament, but none of us won.

Figure 9. Self-portrait of the author in his YMCA wrestling uniform.

Another part of my wrestling activity there was helping, unofficially, the high school team, and referring some of their bouts. The following article was printed in 1938 in the Davenport newspaper:

GLEN HOFFMAN REFEREES BOUTS

Do you remember that little blond referee who officiated at most of Davenport's wrestling matches? He is Glen Hoffman, a student at St. Ambrose.

He began his wrestling career with four seasons on the Lamont high school team, coached by Paul Young, who is now our boys' adviser. Modest and quiet, he doesn't talk about it, but he was rather good in the lighter weights. He now wrestles with the local Y.M.C.A. team and is going to enter the open amateur tournament at Waterloo in early March.

He has refereed six matches for Davenport and when he doesn't referee he likes to sit in the stands and shout with the rest of the spectators.

He is working his way through St. Ambrose College and intends to become a pisciculturist. Do you know what that is?

Besides wrestling, his major hobbies are trout fishing and photography.

Figure 10. The author (crouching, in white shirt) refereeing a wrestling match. Photographer unidentified.

4. University of Iowa 1939-1942

Being short of funds, I stayed out of college for the 1938-39 academic year, working where I could to save money to transfer to the University of Iowa at Iowa City where I began my sophomore year in the fall of 1939 in Liberal Arts with a major in Zoology. Although I was academically a sophomore, because of College rules I was on the freshman wrestling team. I was a contender for the 136 lb weight class, but I dropped out of wrestling toward the end of the season because I couldn't study after strenuous workouts. My grades improved, helping me to obtain a good board (meal) job at the University Hospital near the Quadrangle Dormitory where I lived. It was there that a great kindness helped me. My job was folding linens in the hospital laundry room where the freshly laundered and dried linens were shaken out, liberating much linen dust which greatly aggravated my allergies. The Director of Housekeeping recognized my problem and transferred me to washer of little windows in ward doors and elevators. She said it was a three-hour job for the three daily meals at the hospital cafeteria and if I could do the work in less than three hours, that was OK. So I hurried and completed it in less than three hours. I will never forget her kindness and thoughtfulness. I was not able to recall her name, but an University of Iowa librarian, Denise Anderson, searched it out for me. The kind Director was Winona Ballantyne, who died in 1980 in Illinois.

Wrestling, University of Iowa

In 1939 our University of Iowa wrestling coach was a very nice man, Mike Howard. I was on the freshman team. My wrestling partner was Russell Miller. I found that after strenuous workouts, I couldn't study zoology, my major, very well so I told Coach Howard that I'd have to quit wrestling. He said "why do you do that?" and I told him why. He told me to look at a certain wrestler who was in medical school and on the team. That wrestler flunked out of medical school, but I continued my studies and, after WWII, earned the PhD in Parasitology and Microbiology.

Iowa won the Big Ten wrestling tournament during WWII, and some time [decades] after WWII an Olympic Champion, Dan Gable, became Iowa's wrestling coach. Under Gable, Iowa soon became national champion and the team has continued to compete very well. Now, many years later, Iowa has dominated the Big Ten and National Championships. At this writing they have won the Big Ten and National Championships two years in a row, 2007 and 2008, and many Iowa graduates have become high school and university coaches.

In wrestling, my workout buddy was Russell F. Miller because our weights were similar. After I dropped out, he continued and had a good wrestling record. At 128 pounds he was captain of the first undefeated Iowa U. team in '42-'43. He wrestled for the Big Ten Championship, but lost by only one point. Later he had victories over two Big Ten champions and Iowa was undefeated when he was captain. I knew Russ only in wrestling practice, but he was always a very nice person. In a recent letter, Russ told me of an unusual coincidence. During WWII Russ was a lieutenant in the 517th Regimental Combat team, a parachute infantry regiment, and my sophomore roommate Bill Sherman (deceased), the 118 or 121 pound champion, also a lieutenant, was in a different parachute outfit. With all the soldiers and civilians in Rome, they bumped into each other walking down one of the main streets in Rome!

After the war Russ took his law degree at Drake University in Des Moines, IA, then went into the CIA (Civilian Intelligence Agency) for a 29 year career that included assignments in Denmark, Korea, Pakistan, Thailand, Turkey, and Vietnam. After retirement he worked as a volunteer in the first George Bush's presidential campaign and later received presidential appointments as Deputy Inspector General at the Synthetic Fuels Corporation and then Inspector General at the Federal Emergency Management Agency (FEMA). He spent four years on the latter during which he worked on the recovery from hurricane Andrew which slammed into south Florida. I think he had an excellent career and I also like to think that wrestling helped both of us to get started.

Zoology Studies, University of Iowa

During my undergraduate studies, zoological courses were most important to me because of my interest in fish and fishing. Organic chemistry also was related, as was geology because of fossils, including fossil fish and snails. Learning to prepare mounted fish was also re-

Figure 11. The author at work painting a plaster cast of a trout. Self-portrait by the author.

lavant, and the Iowa State Conservation Department hired me, part-time, to mount some for them. The usual taxodermatist skins the fish and stuffs the skin. It must be an art. But Butch Mues, my teacher in Museum Technique, taught me to make plaster casts of the fish with attached paraffin fins, and then paint the whole. Butch's were very nice, very real looking and straight, rather than curved like most taxi-dermatists make. Mine were also straight. One such mounted fish, a nice smallmouth bass, is still in possession of my son, Lyle, nearby. But most notable, I think, was my large spoonbill catfish that was assigned to the Backbone State Fish Hatchery where I had worked, and where

Figure 12. Smallmouth bass mounted by the author and now in the possession of his son. Note one drawback of this method of mounting: the dorsal and lateral fins have suffered damage in several moves. Photo by Lyle Hoffman.

my brother, Melvin, was working. It hung there for many years until someone stole it!

The most important factor in my undergraduate education, however, was the help and enthusiasm of Professor Nolf, which started with my enrolling in his general parasitology course. His enthusiasm and friendship was the start of a wonderful relationship that would last the rest of his life and much of mine. This friendship was not confined to the classroom; it extended to his family as well. His wife, Gladys, and two sons, Bruce and Richard, were part of that wonderful relationship. I don't recall how it came about, but I took younger son, Richard, fishing at a small stream near Iowa City. We were using earth worms for bait and Richard caught a huge carp, so large that I had to help him land it. Many years later he said it was the largest fish he ever caught. Also, even more years later, when I was living alone in my West Virginia home after my wife Carolyn's death, Rich, as he was known by that time, went out of his way on a trip from his home in St. Joseph, MO, to Baltimore, MD, to stop and thank me again for helping him catch that large carp so many years ago! I am forever grateful.

Professor Nolf's other son, Bruce, had a good career in geology in Oregon. Rich also had a good career as Director of the St. Joseph Museum in St. Joseph, MO.

In addition to my feeling to be part of the Nolf family, I was part of a larger family. Dr. Nolf's major professor was "Uncle Jimmy" of the University of Kansas at Lawrence, another very helpful, friendly person, so being Dr. Nolf's student I was part of Uncle Jimmy's professional family.

Reserve Officers Training (ROTC)

I really wasn't interested in ROTC, but, at the University of Iowa, it was required in men's freshman and sophomore years so I was enrolled in ROTC in my sophomore year. I don't remember much about it except one "highlight." At marching drill one day, Lenny, a housemate of mine, and I were in parade drill when Lenny simply started walking off the field. Our leader, an Army Major, barked "Where the hell are you going?" in a rough voice. Lenny turned and said simply "I gotta go" and believe it or not, the gruff major smiled and said no more!

5. College Friends and Roommates

One of my Quadrangle Dormitory mates was Richard (Dick) Spencer (deceased). He was very talented and invented the character, Herky the Hawk, symbol of the University of Iowa Hawkeyes, which is still used, probably with many modifications. He was an art major, art editor of the art publication Frivol, cheerleader and member of ROTC Engineering Unit. During WWII he was a captain in the paratroops in Italy, France, Belgium and Germany and received three purple hearts. His latest work was editor and finally publisher of the *Western Horseman* of Colorado Springs, CO. Dick was very well liked. We miss him.

Another good friend in the Quadrangle Dormitory was James (Jim) Platt. Jim was an archery enthusiast. He was good at it and careful, but shooting an arrow over the heads of classmates was a little scary. Fortunately no one was ever hurt. After the war he returned to the University of Iowa for his Master Degree. During this period Jim invited me to go squirrel hunting with him. He wanted to shoot a cat we saw in the woods, but I convinced him that it was probably someone's pet. He didn't shoot it—I was proud of him. Afterward he did editorial work for the Des Moines, IA, *Register*, then for the *Idaho Statesman* in Boise, ID, and finally he was a special projects writer for the Office of Secretary of Defense in Washington, D.C. Jim died of complications of two strokes at home in Alexandria, VA, in 2005.

Roommate Bill Young (deceased)

In 1937 while attending St. Ambrose College in Davenport, IA, I was living with my benefactors, Paul and Lyonne Young. Their nephew William "Bill" from Manchester, IA, was similarly a guest so we shared a nice room and had the run of the house and other kind benefits. Bill was also working as office boy for a cigarette company. We were very compatible, so it was a very comfortable situation for me. In addition to the use of a room we shared in the family meals at a very fair fee. We sent our laundry home for washing which was the custom then for college students. One of Bill's brothers was a weekday guest also. He worked at a farm machine company in Moline, IL, just across the river.

Another brother, Dean, was sometimes a similar guest. Both brothers preceded Bill in death. In later years, Bill and wife Lorraine named a daughter, Deana, in memory of Dean.

Sometime later Bill, being innovative, devised a system of recovering spilled grain on the highways near where they lived in Clarion, IA. Bill died of heart complications in 2001. We miss him greatly.

Roommate Bill Sherman (deceased)

Bill was my sophomore roommate at the University of Iowa, Iowa City, IA, 1939-40. We were picked for roommates because Bill, also a sophomore, was a wrestling champion and I had agreed, although a sophomore, to be on the freshman wrestling team (as was the custom then for transfers). My wrestling must have been OK because Coach Mike Howard awarded me a freshman wrestling sweater. Although Bill was about ten pounds lighter, he usually beat me because he knew more maneuvers. We were very compatible. I recall that I was doing some "moonlighting" photography for pay and had to do the film developing in our room, late at night. Bill would go to bed at his regular time and he never complained about my late night activities or the lights when I had to turn them on.

We lived in the Quadrangle Dormitory. It was a real quadrangle with a courtyard in the center. When I was at home and Bill was coming home, he would let out a real Tarzan yell (ah ee ah!) and I would go to the window and answer in kind. That way he would know if I were home.

Also, very interestingly, when Bill saw a pretty girl on the classroom part of campus, he would let out a very realistic horse whinny! Surprisingly, I never heard that any of them objected! Bill was a good guy.

Bill was a paratrooper during WWII. Later he became a school teacher and was killed by colon cancer in 1982. Ironically, our coach Mike Howard also died of colon cancer.

Roommate Roland (Steb) Stebbins

Roland was my roommate when I was a junior, he a freshman. He also was a very considerate roommate. I can't recall what he had intended to major in, but he dropped out after one year. I left a note on

Figure 13. The author (left) meeting new roommate Roland Stebbins in 1940. Photo by the author.

my door for him once and signed it GH. From then on my neighbors in the dormitory, The Quadrangle, called me "GH."

I think Steb didn't need a college degree because he was a good mixer and manager type. He became head of a city group in Rockford, IL, and did very well. One sad note: His younger brother, Dwayne, died early of Alzheimers. At 87 Steb is doing well, but slowed down by macular degeneration of the eyes. His good wife, Sheila, is a big help for him.

COLLEGE FRIENDS AND ROOMMATES

Roommate William (Bill) Leaming

Bill was my roommate for my senior year, 1941-42. He was a junior and his senior year was delayed by WWII, but after the war he finished his BA and completed an MA in Social and Economic Science. I think he became a bank vice president, but he didn't respond to my letter to him in 1991.

His father worked for a washing machine company, was handy with tools making, e.g., good hunting knives out of discarded saw blades (good metal) with nice aluminum handles. I still have the one that Bill gave me. It is a good, durable knife that holds its edge well.

6. My Brother, Melvin

All Mel's life, hunting was probably his greatest joy and motivation. It may have started when he was about six years old. His Christmas present that year was his first gun, a BB gun. He found a rabbit sitting in the machine shed and shot it, probably accidentally in the eye, killing it. We ate it, of course. I think he was a hunter from then on. Later he was a better cow milker than I, so after supper when he and Dad milked the cows, I was often detailed to pop corn for our evening snack.

In high school we were both on the wrestling team and occasionally we were paired in practice. He was a year older and heavier. I remember that he pinned me at first, but after that I learned more maneuvers and beat him so, as we said, "we took turns winning."

After high school Mel worked as an apprentice carpenter for our neighbor, Ben Kotek. He worked at that for several years and never forgot how to use woodworking tools well.

Figure 14. My brother Melvin in 1938. Photo by the author.

Figure 15. Bob Cooper and Mel seining trout in 1938. Photo by the author.

When I left, about 1938, to begin college, Mel was able to start his lifelong career of fish culture at the State Trout Hatchery in Backbone State Park where we had worked together until I left for college. I recall one bad *faux pas* we made. We treated one small pond for a fish disease and left the chemical in too long, killing all the nice trout. We buried them nearby and the superintendent, Robert Cooper, treated us kindly.

Years later my brother often commented that Bob Cooper was the best boss he ever had.

There was a large sturgeon named "Oscar," kept in one of the larger ponds, that Melvin took to the State Fair in Des Moines, IA, for many years.

Melvin was innovative. At one point he installed four pay trout pellet dispensers at some of the ponds. They were very successful. At another time he invented a powered cement pond bottom scrubber that worked well for many years.

He was also very kind. For example, when I was a senior in college I borrowed money from Mel to buy a used car. Five years later, after

WWII, I repaid him, but he refused to accept the interest. At the time I left home to start college, I didn't know what to do with my caged raccoons that I was raising for their pelts. Mel kindly took them over so I didn't have to worry about them or the cages. Also, during my year out of college, between my freshman and sophomore years, when I didn't have a car, he let me use his brand new Pontiac on dates. Not every brother would do that!

After he retired, he built a retirement home for himself and his wife, Eileen, using help only on things he couldn't do alone. In addition to the house, he made many things of wood—cedar chests, inlaid bowls, fish on nice plaques, inlaid table lamps, etc., etc.

He was addicted to nicotine, couldn't quit smoking, and complications from smoking killed him in year 2001 at age 84. Surviving is his schoolteacher wife, Eileen, daughter Linda and her husband Tom Kringlen and grandchildren Michelle, Kersten and Ryan. There are also two great grandchildren, offspring of Michelle and Joe Ludwig.

Mel's wife Eileen and her sister Ersileen, now Marquardt, are identical twins. At least once when Melvin and Howard Marquardt were dating the twins, they switched partners on the boys. Later Mel and Howard learned how to tell the difference. The relationship among the two pairs was so great that they had a double wedding, and the four honeymooned together.

In later life Mel, as family called him, became fond of jokes. I recall two of his favorites: "Did you hear about the kidnapping in Strawberry Point? There was a kid napping on the porch and his mother caught him!" Also, "Do you like to hunt squirrels? I can tell you how to catch one—climb a tree and sit on a limb. The squirrel will think you're a nut and he'll catch you!"

7. World War II

After graduation and before my enlistment in the U.S. Army in the fall of 1942, I worked as a biologist for the Iowa State Conservation Commission with headquarters at the Lansing, IA, regional center. Most of the work was sampling fish populations in trout and bass streams. My leader was a veteran fisheries worker who had learned to use an engineer's transit to map the streams we sampled. We also did some routine water chemistry. When we were working out of the Lansing office I could flyfish on the edge of the Mississippi River after work. I did catch some fish and the kind cook at the little restaurant where I ate cooked them for me. She, in friendly fashion, called me the "bugologist"!

I joined the U.S. Army in the fall of 1942. I was shipped to the Army Medical Station at Camp Crowder, MO, and assigned to the medical lab as a lab technician where I did the usual bacteriology and basic blood tests. Experiences in this lab were generally useful to me in later work because up to this point I had very little experience in bacteriology. A kind civilian employee bacteriologist, Alice, quickly taught

Figure 16. The bacteriology lab at Camp Crowder in 1944. Photo by the author.

Figure 17. Bob Rostberg at Camp Crowder in 1943. Photo by the author.

me the basics of medical bacteriology. It is unfortunate that I can't recall her last name. My closest buddy was Robert Rostberg (deceased) from Louisville, KY, a quiet intelligent young man. He worked in that lab until he was discharged after the war and returned to the paint company he had worked in before the war. He died rather early. Bill Purcell, a high school teacher from Pottstown, PA, was another close friend. He returned to teaching, but died too young. Henry Wallbrun, with a BS from the University of Chicago, was very intelligent. He returned to the university to take a PhD in genetics and worked at the University of Florida at Gainesville where he retired. After retirement he created new varieties of orchids. He had a very serious stroke about year 2000 and I haven't heard from him since 2002.

I recall two unhappy incidents while at Camp Crowder. A young medical doctor (Lieutenant I think) prescribed a too high dose of sulfa for a young man very ill with meningitis. The young man's urethras became plugged with sulfa crystals and he almost died. The other incident was when a soldier attempted suicide by drinking a potentially fa-

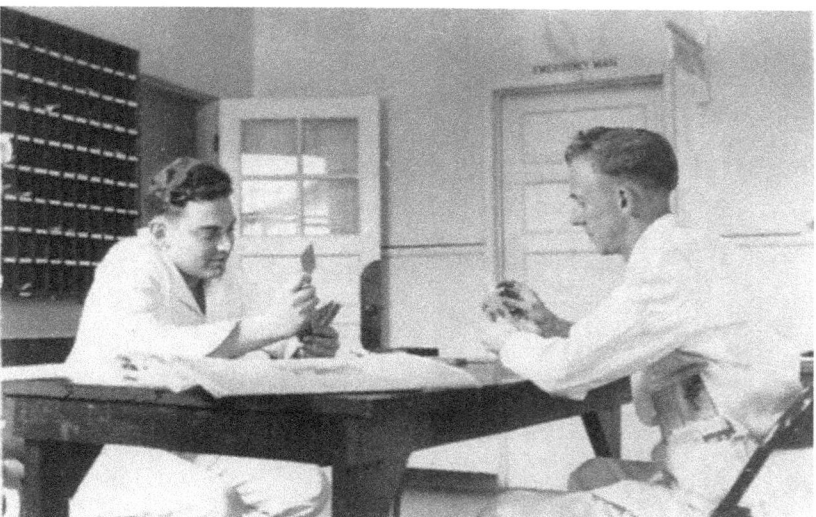

Figure 18. Henry Wallbrun and Bill Purcell at Camp Crowder in 1944. Photo by the author.

tal chemical. A young medical corps lieutenant, not a physician, was instructed to come to our lab to obtain a stomach sample jar. We had two kinds, one sterile, one not. When I asked him which he wanted he, not knowing which he needed, became angry. Because of that I, being only an enlisted man, was penalized by having to dig a hole in the ground and then fill it up! After the war Camp Crowder was demolished and became a housing development.

In September, 1944, I was assigned to the 361st Medical Composite Lab as sergeant and we were shipped to Le Havre, Normandie, France. There were 26 of us in that composite lab. Unfortunately our lab equipment was on a different ship which went to England. Our equipment, as intended, was transferred to Salisbury, England, where we were supposed to establish a central lab. Apparently another lab composite established the lab and the original 361st lab was transferred there much later. With all the Army activity in that area, it is not surprising that there would be mix-ups.

We were moved to an otherwise unoccupied apple orchard, not far from Le Havre, where we were to camp for two weeks. We paired up and slept in pup tents, two men to a tent. Our leader, Colonel Aron-

Figure 19. Sgt. Alderman and the author at the quarters we shared in the Normandie apple orchard. Photo by the author.

son, managed to find a mess tent, the fall weather was nice, and, with no duties, we enjoyed the Normandie apple orchard and surroundings. I recall hiking along a nice trout stream and seeing brown trout and European salamanders. The farmers in the area were friendly, particularly the children.

The veterinarian assigned to our unit was very resourceful and managed to acquire good food for us which we cooked or warmed over a campfire. On one occasion he supplied delicious canned turkey. Sometimes there was singing around the campfire. One song I recall was *Ave Maria* sung very well by H. Bryant (I've forgotten his rank). He had been in a church choir. I tried to contact him recently to no avail, probably because of the Privacy Act.

Our commanding officer, the late Dr. (Lt. Col.) Joseph D. Aronson, who had served in World War I, was able to get us transferred to the Pasteur Institute in Paris. The second in command was Major Kushner, a histopathologist, and a kind man. The kind French staff

WORLD WAR II

Figure 20. A page from my wartime photo journal, from our time in Normandie. You can't take the naturalist out of the soldier!

was able to clear out three rooms for us, but I don't recall where our equipment came from. I was a sergeant, but during our stay there I received a 2nd lieutenant commission from General Dwight Eisenhower based on an application I had made about a year before. I don't think the promotion would have been possible without the help of my parasitology professor, the late L.O. Nolf. It was so characteristic of him to help students that I here include a copy of that letter:

The State University of Iowa
Iowa City, Iowa
Department of Zoology

February 24, 1944

To Whom It May Concern:

    Pfc. Glenn Hoffman, 37440250--Det. Med. Dept., Camp Crowder, Missouri--has requested me to write a letter to accompany his application for a commission in the Sanitary Corps of the A.U.S.

    I have been informed by the Headquarters Omaha Office Procurement District that "the army is very much in need of parasitologists," and since Private Hoffman has had training as a parasitologist, I believe he would be of greater service to the army in this capacity.

    I have a very high regard for Hoffman's ability. As an undergraduate, he took graduate courses and his work was equal and frequently superior to that of graduate students. He conducted research for the Department of Fisheries of the Iowa Conservation Commission. He also pursued a problem here at the University of Iowa with such vigor and success that I have recommended him to the Wisconsin Conservation Department as a research man, after the war, to participate in the survey of the parasites of fish in Wisconsin, now being conducted in the summer months of Dr. Bangham and myself.

    Hoffman has profited considerably by his fourteen months experience in the diagnostic laboratory at Camp Crowder and has learned to appreciate the problems pertinent to its operation.

    A number of my former students are officers in the Sanitary Corps and I believe that Hoffman is worthy of joining their ranks.

Sincerely yours,
L.O. Nolf
Assist. Prof. of Zoology,
Hygiene and Public Health

LON:EB

Much later, in 2005, I met a nice lady, Dorothy Gafford (deceased) who had served in General Eisenhower's secretarial pool in London and Paris. Because she spoke so highly of him, and I was pleased to receive the promotion, I include a copy of his promotion order:

CONFIDENTIAL

9 November 1944

AG 201 Hoffman, Glenn L. MPAB
HQ EUROPEAN T OF OPNS, USA
APO 887
COMMANDING GENERAL
SEINE SECTION, COM Z,
APO 574, U.S. ARMY.

COURIER

GLENN LYLE HOFFMAN, 37440250, PFC, MEMBER 361ST MEDICAL
COMPOSITE DETACHMENT LABORATORY, APTD 2D LT SN C - AUS
BY WD. ASN 0928275. DISCH AS EM IMMEDIATELY. FOLLOWING
DAY ADMINISTER OATH OF OFFICE AS 2D LT SN C - AUS. -
NOTIFY THIS HQ DATES ACCOMPLISHED SO THAT CONFIRMING
AD ORDERS MAY BE ISSUED. EFFECTIVE ON DAY AFTER DISCH
DIRECT OFFICER TO PROCEED TO UNIT WITHIN YOUR COMD WHERE
APPRORIATE VACANCY EXISTS. IF NONE AVAILABLE, DIRECT HIM
TO REMAIN WITH PRESENT UNIT UNTIL REASSIGNED BY THIS HQ.
IN EITHER CASE, ADVISE THIS HQ IN REPLY ACTION TAKEN RE
HIS ASGMT SO THAT SAME MAY BE CONFIRMED IN AD ORDERS. WD
MAILING LTR OF APMT. ACCOMPLISH AND FORWARD FORMS LISTED
PAR 3b (4) CIR 59, THIS HQ, CS.

EISENHOWER

OFFICIAL:
RENE D. DANIELS
Captain, AGD
Assistant Adjutant General

TRUE COPY:
RALPH A CONNELLY
2d Lt, MAC

Figure 21. Our lab in the Pasteur Institute in late 1944. Photo by the author.

Because of the promotion, I had to move out of the enlisted men's quarters and to a room in a nearby family's apartment. The family was very kind to me and we became lifetime friends. I was not married and during my Paris sojourn I met two nice young ladies and arranged to take one of them to a movie later. During the movie she showed me her hands, all swollen from chilblain because she had to work in an unheated office. Also, she told me her boy friend or fiancé had been killed in the war. But later, after I was transferred to near Oxford, England, her girl friend wrote to me that she had committed suicide. War is HELL.

On one occasion, I met another nice young lady in a street car in Paris. We became acquainted and I took her to a nice restaurant where there was a string trio musician group. I asked them if they could play Massenet's *Meditation for Thais* for us. They did, without sheet music before them — I was impressed. On a later occasion the fine young lady asked me to dinner at their apartment. She asked what I would like for dinner and I chose duck. In spite of all the shortages during the war, her mother managed to go into the countryside to purchase a duck, and we had a fine duck dinner. All over Europe, I always found

Figure 22. The author and another officer (unidentified) with the family Coutant who provided living quarters during our stay in Paris. Photo by the author.

friendly people, even in Germany. Apparently they were grateful that we Americans had helped them.

Our lab was transferred to Hopital Beaujon, a large former French hospital, but I recall only two incidents from that location. On one occasion we had to carry in wounded German soldiers who were very heavy, but very young. One was very frightened and my attempts to soothe him with my inadequate German failed. The other was very disconcerting – at reveille one morning, the top sergeant was absent, but when he did come, he lied about why he was late. I think he overslept. Colonel Aronson was very angry and said he would court martial the sergeant! But before a court martial, there must be a hearing of the officers of the unit. The colonel was the only one who voted for court martial, so it was dropped. Good.

Before Germany surrendered, my brother, Melvin, was in an anti-tank tank unit close to the battle of the bulge. At one point, his officer ordered him to shoot the young German soldiers huddled behind a haystack. My brother rounded them up and marched them to his officer, refusing to shoot them. The officer received a medal, my brother nothing. My brother suffered frost-bitten feet because soldiers in his

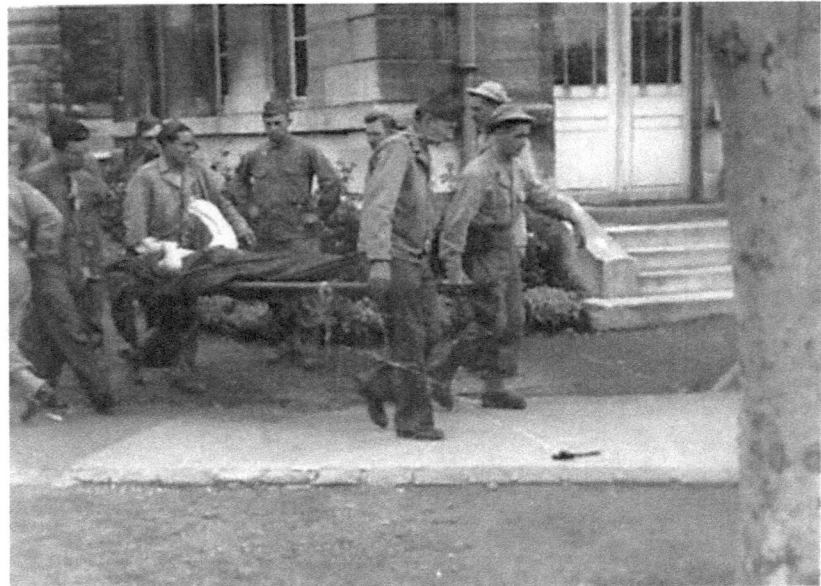

Figure 23. A wounded soldier being carried into l'Hopital Beaujon by some of my buddies. Photo by the author.

unit had been issued only leather boots and they got wet feet in the wet soft snow. Fortunately, Melvin survived.

There was no position for a new officer in the Paris lab, the 361st Medical Composite Detachment Lab, so I was transferred to the 318th Station Hospital Lab near Oxford, England, as Assistant Lab Chief in January, 1945.

My tour of duty with the 361st impressed me very much because it was my first ever overseas trip and the beginning of an Army medical lab. It was also an opportunity to make new friends under rather unusual circumstances.

In the new location, near Oxford, we slept in tents, but the lab, proper, was in an adequate wooden building. The weather was usually nice so my chief, Captain Loyal Suter (deceased), and I bought used bicycles and had nice bike tours in the pleasant countryside, including a tour path along a scenic canal. Oxford University, nearby, allowed me to use their library so I brushed up on some fish parasitology. I also tried a one-person narrow skiff, I think it was called, on

Figure 24. L'Hopital Beaujon while we were stationed there in 1944. Photo by the author.

the Thames River. I was warned about the danger of tipping over, but I didn't tip.

I was a parasitologist lieutenant in the Sanitary Corps (now part of the Medical Corps). One day at our medical complex near Oxford, I was asked to check the filter bed of the sewage system. It was rather small, looked OK, and I sampled some of the microscopic organisms which looked OK to me. I couldn't think of anything else to do so I told the sewerage caretaker to stir up the rocks and sand to make it work better. He very politely did not, because, as I learned later, it is not wise to disrupt the biological "cleaning" activities in

Figure 25. Our quarters at the 318th Station Hospital. Mine was the first tent on the right. Photo by the author.

such a sewage filter. It really wasn't my job, so live and learn. Sometime later I toured some large sewage plants and saw how the various parts worked.

The hospital lab station near Oxford was a pleasant tour of duty for me. In June 1945, I was transferred to the 49th Station Hospital near Cambridge, England, as lab chief. I still had my bicycle so again enjoyed British countryside. During this tenure, Germany surrendered. That night I was officer on duty when a young soldier, very happy, went down the row of hospital tents beating loudly on a tin pan. Such was not allowed, but I didn't have the heart to stop him on such a wonderful occasion.

The most unusual coincidence of my life occurred to me in Oxford, England, and London, England. While stationed near Oxford in 1945, I had one date with a very pretty girl. Soon after that I was transferred to the continent and about a year later was in London on leave. I was walking down a sort of secondary street, in the middle of London where there were thousands of people, and there SHE was walking down that same street. She had no available time so we didn't have a second date. I wasn't married then.

Figure 26. Capt. Suter in the 318th Station Hospital Lab in 1945. Photo by the author.

On this same trip, I went to Royal Albert Hall to hear Yehudi Menuhin, the famous violinist, but no seats were available. However, a kind English lady recognized my plight and gave me a ticket to the performance. During war time this was a real treat. Also at a restaurant, the owner, recognizing my U.S. uniform, gave me a large chocolate candy bar in spite of British shortages of such.

In December 1945, I was transferred to the 7th Medical Lab at Gräfelfling, near Munich, Germany, to assist in the influenza section under Lt. Victor Sprague on a Dr. Jonas Salk influenza project. There were German employees in that station and they were very kind and helpful. I recall that one made a photo print box and a very nice leather lab tool case for me. Later Captain Sprague PhD became a world famous Microsporidia protozoologist.

Shortly before I was transferred to the Gräfelfling lab, the young sergeant, who sort of became my sergeant, was called upon to embalm General Patton who had been killed, while in his jeep, by a drunken American soldier in a big truck. Alcohol often causes evil! The sergeant was a very nice young man. I wish I could remember his name.

Figure 27. The author at work in the influenza lab in 1946. Photo by the author.

Next I was transferred to an Army lab at Darmstadt, Germany. I wasn't there very long, but I recall the nice young German boy who picked up our laundry for his mother to wash. One day he and I went for a walk. He showed me the front of a nice church with a steeple, and said, "now look at the back"—it was completely bombed out. He said, "You Americans did that!" But he seemed to hold no grudge because the war was over.

In August 1945 I was transferred to the 228th Station Hospital lab in Liege, Belgium, which became the 91st General Hospital. Captain

Figure 28. Ernst (surname unknown) outside the lab in Darmstadt. Photo by the author.

Eugene Kennedy, bacteriologist, was my lab chief. This lab was on a rather high hill in a compound called The Citadel which had been a Belgian hospital and/or fort. It was nice to be in regular buildings again. The Belgian people were kind.

In October 1945, on a post-surrender educational leave, I visited the Marine Biological Lab at Plymouth, England, where I met the late Dr. Nora Sproston who helped me with parasite preparations. I was also able to visit the U.K. Freshwater Biological Assn. Lab on Lake Windermere, northern England, where we had a nice hike in the mountains and the lab staff prepared a fancy large trout bake. The trout, a brown (*Salmo fario*) was one of those removed from the lake because they ate too many smaller fish.

Figure 29. The Marine Biology Lab near Plymouth, England. Half of the building was blown away, but the standing half remained in operation. Photo by the author.

At the Liege hospital/lab, my roommate and friend was a captain urologist, but I can't recall his name. We went on outings together and had many enjoyable moments. When the Japanese surrendered he said, "Glenn, let's celebrate," by which he meant to drink his bottle of congac (alcoholic drink). At that time I was neither a drinker nor a teetotaler. After celebrating, on the way to dinner at the Army dining room, I realized that I couldn't walk a straight line. I vowed to myself never to drink again. And I didn't except for sipping champagne at my reunion with my Parisian landlady, many years later.

In Liege I met a girl, Mary Hubo, who spoke English well and was a nice person. I recall taking her to our Army dinner once, but I think she was so overwhelmed by all the English-speaking officers that it ruined her appetite. I never heard from her again although I mailed her my leftover Belgian money.

My last Army tour of duty was a short one in Antwerp, Belgium, a sort of staging camp for preparation of those to return soon to the United States. Because my title was lieutenant in the Sanitary Corps

Figure 30. Self-portrait of the author overlooking Lake Windermere.

(should have been Medical Corps) the only duty there that I recall was for me to inspect the mess (dining) hall. The house flies were rather numerous but all I could do was to suggest more fly control, a minor matter.

Arriving back in USA in March, 1946, I was stationed for a few days at an Army staging camp near New York City. My only memory of that time was running across a former Camp Crowder, MO, lab co-worker, named Mickey. I took her to a play in NYC. She was a very well-liked person. Then I was transferred to a Separation Center, Camp Grant, IL, where I received the Honorable Discharge dated May, 1946.

During the war, Uncle Frank Hoffman (deceased) had kindly stored my 1942 Pontiac at his place. After the war I put in a new battery and gas, but it wouldn't start. Dad helped me haul it to a local repair shop.

The owner, a neighbor, said the car was OK. Being used to the Army jeep controls, I had pushed the wrong button! The joke was on me.

The funniest war story I recall was that of our beloved zoology teacher, Dr. Robert King, who was in the Army during WWI; he worked in a medical lab. He said, "One day in Paris we were to transport some stool (fecal) samples, but we ran out of containers so we used candy boxes—but someone stole them!"

Toward the end of the war a white soldier insulted a black soldier. The black chased the white one with a razor. The black was being court-marshaled, not the white. I was a member of the court, and I asked why the white soldier was not being punished also. I was told that that was not part of the proceedings.

8. University of Iowa, PhD Program 1946-1950

I was discharged from the Army in May, 1946, and wanted to start graduate school under Professor L.O. Nolf at the University of Iowa. However, it was not possible to start in mid-semester, so Prof. and Mrs. Nolf (both now deceased) kindly took me in, practically as a family member, until I could enroll in summer school. During the rest of the semester I worked unofficially in Prof. Nolf's lab. Their kindness will never be forgotten. In that same vein, after I was married, they provided Carolyn and me with a garden plot in their garden.

In 1946, post WWII, college accommodations were still crowded due to Army or Navy programs so four of us were crowded into one room in the Quadrangle Dormitory, the same dormitory I lived in be-

Figure 31. The author with Dr. and Mrs. Nolf at their retirement home in 1984. Detail from a photo by Carolyn.

Figure 32. The Woodruff lab of the Wisconsin Conservation Department in 1946. Photo by the author.

fore WWII. I've lost track of my roommates, but I recall that Paul Lang moved to North Dakota; Vito Lopin (deceased), a physical education MS, moved to Rockford, IL; one moved to South Dakota; and one was from the Amana colonies, nearby. In 1948 I moved out to get married.

My PhD degree would be in Zoology, specialty Parasitology, with a minor in Microbiology including Bacteriology and Mycology. In the course of completing my minor, I published my first scientific paper, "Isolation of Saprolignia and Achlya with penicillin-streptomycin and attempts to infect fish." (1949, *Prog. Fish Cult.* 11: 171-174). Prof. G.W. Martin was my teacher.

During the four-year PhD program I was able to spend three summers on a fish parasite program of the Wisconsin Conservation Department in a lab near Woodruff, northern Wisconsin which resulted in the experimental basis of my PhD thesis and a 1958 publication, "Experimental studies on the cercaria and metacercaria of a strigeoid trematode, Posthodiplostomum minimum." (*Epp. Parasitol.* 7: 23-50). In this study I demonstrated two subspecies using host specificity. I've often thought it would be interesting to examine the DNA from both sub-species. My lab mates from University of Iowa were Professor L.O.

Nolf, leader (a good one), and Tom Parsons who was a professor at Grinnell College also. Tom was a very nice person and a good canoeist who had married the daughter of our anatomy professor, Dr. Stromasten. Tom died very early due to a massive heart attack. The graduate students from the University of Wisconsin were Ed Schlueter and Ralph Duxbury. Ed is now Professor Emeritus at the University of Texas at Denton, TX. Ralph retired from the Army Medical Center, Washington, DC. Dr. Frank Brooks (deceased), from Cornell College, IA, was also at the Woodruff lab, and helped me with trematode cercariae. He was also my cabin mate, and a good one, for one summer. Professor Chester Herrick, University of Wisconsin, was a frequent visitor who was good at untangling fishing lines. He also died of a massive heart attack which happened when he was driving to work; he pulled over to the side to avoid endangering others. He was a very nice person.

The Movie, Save Those Fish

The major part of the Wisconsin Fish Parasite summer research concerned sampling fish from Lake Arbor Vitae, near the lab, digesting the fish in pepsin fluid, and counting the various surviving trematode parasite larvae. To do this, Professor Nolf had constructed an ingenious shaker apparatus in a warm water bath to digest the individual fish in jars of digest fluid. This was to determine if the fish had fewer parasites after the sampling areas had been treated with copper sulfate/copper carbonate to kill the intermediate hosts—snails. So samples were taken before and one year after reducing the snail population.

A movie was made, showing the collection of fish, treatment of the chosen areas with the copper sulfate/copper carbonate solution, the fish digest operation, and Professor Brooks studying the larval parasites from the snails. The movie began with a scene of a fisherman (me) flyfishing from a canoe and actually catching a crappie. The movie ended with two fishermen cooking fish over a campfire, eating them, and indicating that they were good. A copy of the movie, *Save Those Fish*, is stored in the library at the Fish Health Lab, Science Center, U.S. Geological Survey, 11649 Leetown Rd., Kearneysville, WV 25430.

Weekend Fishing Trips

During the summer sessions at the Woodruff, WI, lab we had many weekend fishing trips. Most were good, but I remember one bad one. I drove alone to a small flowage for a one-man canoe float, looking for muskies, but left my fishing tackle at the cottage. It was several miles away so I had to return to get my tackle, no fun. It is not only the elderly that can be forgetful! And I don't recall catching any fish on that trip.

One day three of us went to a small lake to fish. We were walking on a trail with brushy growth on both sides, as well as a steep hill on one side. A skunk was on the trail, walking toward us! It wasn't easy but we managed to leave the trail in time—the skunk won the right of way!

There was a moderately good trout stream not too far away and I found a nice pool where I could usually catch two or three nice brown trout on small nymph flies. I recall showing one physician how it was done. He had never flyfished.

And, of course fish are unpredictable some times. One time several of us went to an artificial lake where fishing had been reported to be good. There was a boatful plus my thirteen foot aluminum canoe with me and another student. We hadn't fished very long when my helgramite-baited hook was taken by a nice crappie about ten inches long. I landed that one and four more about the same size, all using that same helgramite. No one else, neither those in the other boat nor my companion, caught anything although all were experienced fishermen. The helgramite bait might have resembled something those fish were familiar with. It was fun. Those were wonderful days.

Back at the University of Iowa and the Mentorship of Dr. Nolf

Dr. Nolf taught me how to handle the rather voluminous literature on fish parasites which was mostly scattered in several journals. He showed me how to make a fish-parasite file which I maintained on 3x5 cards throughout my career of 39 years. At one point I told him that I intended to condense it all into one book. He said "You'd better wait twenty years." He was right; the first edition of my book *Parasites of North American Freshwater Fishes*, was not published until 1967! Some years later, at a dinner, I overheard him tell someone that Glenn is the best fish parasitologist in North America. It was generous of him. His major professor was "Uncle" Jimmy Ackert at the University at Lawrence, KS—so we had a family line—Uncle Jimmy to Owen Nolf to Glenn Hoffman and many others. My lab mate at the University of Iowa was Richard Stoner (deceased) who worked at the large government lab at Brookhaven (Long Island), NY.

Working in the same lab, under Dr. Nolf, was John Ohlsen who had been a lab tech in the U.S. Army during WWII. He received the MS in Zoology, specialty Parasitology, and started a medical lab in Hastings, NE, which he maintained until his death in 2006. He was a good friend.

Dr. Nolf was a very kind and considerate person. For example, one time he had to be away and asked me to give his exam in a conservation course. He also asked me to grade it, which I did. I misgraded the question on the method to treat a frozen fish pond in winter. The answer was the results of a study using charcoal to hasten thawing. I had missed that report and graded that question wrongly. Dr. Nolf kindly corrected the mistake with no verbal punishment to me and explained to the students why I had goofed.

Another example of Dr. Nolf's friendship to students occurred when I gave my first seminar talk to the faculty and students of the zoology department. I had started my thesis project on the life cycle of a parasite helminth (worm) at the Wisconsin field lab three years earlier. I continued that project in the zoology lab at the University of Iowa. I introduced my talk, explaining that Dr. Nolf was my major professor and then said "I have been working with this worm for three years." The audience, as well as Dr. Nolf, took it as a good slip-of-the-tongue joke. Someone declared that it was better than a prepared joke!

Dr. Nolf often had a good joke or corollary to fit the situation. He and I were both rather small, and once when I complained about it, he told me not to worry because when a tall man comes to an ordinary fence he will step over it and we shorter ones can go under—and we both get to the same place! Dr. Nolf always enjoyed a good story, funny or not, and whether he or another told the story, he ended it with "That's a corker."

There is often controversy over the harm foxes do to other wildlife. He recalled a publication about someone testing that. The researcher had two nearly identical enclosed plots. In one he put foxes and rab-

Figure 33. Dr. L. Owen Nolf in 1967. Photo by the author.

bits; in the other he put rabbits only. At the end of the experiment there were nearly the same number of rabbits in each.

We didn't discuss religion very much, but one time I took him to one of my favorite trout fishing streams in northeast Iowa. We parked on a hill and walked down into the lovely trout stream valley and he remarked, "This is my idea of religion." I've often recalled that moment.

He never forgot a student. A couple of years after graduation, my good lab mate, Richard Stoner, I and Dr. Nolf met at an annual meeting of the American Society of Parasitology. Dr. Nolf took us both to the musical, *South Pacific*, a great treat.

In another division of the Zoology Dept was Chester (Chet) Roys whom I first met in the Army at Camp Crowder, MO, during the war. He was in the Sanitary Corps and gave a lecture to us on one occasion. After the war, Chet, whose parents lived in Chicago, returned to the Zoology Department as I did. I recall riding with him in his car to an American Association of Sciences meeting in his home city, Chicago. He finished his PhD and taught at Tufts College, Medford, MA. We kept in touch until 1993 when, I think, he passed away. He was a good friend. I think he was a good teacher, and, among other things, took his zoology students on field trips to the Caribbean Islands, a treat for students.

The final phase of my graduate program was defending my thesis concerning the life cycle of the white grub, *Posthodiplostomum minimum*. Part of the work concerned freezing the parasite from its cyst in fish tissue using pepsin-trypsin digest. The head of the department, a physiologist, pushed me on the chemical technicalities which I defended thoroughly because I had used and studied it. So I defended successfully and Dr. Nolf remarked to me later, "You aced the department head!"

During this period I gave my first scientific presentation at the University of Missouri, Columbia, MO. The report was on my research with *Posthodiplostomum minimum*. I attended the meeting with my major professor, Dr. L.O. Nolf. Coincidentally, my granddaughter, Robin Craig, is a student there now (year 2007).

Professor Nolf was always a very kind, considerate and helpful teacher. He died in 1991 of prostate cancer.

9. Carolyn Elise Wilson Hoffman

As important as all the events of the preceding chapter were to my professional development, the main event of this period was my getting acquainted with Carolyn Elise Wilson, a MS candidate in Zoology. One Saturday in 1947 when Notre Dame was coming to Iowa City to beat the University of Iowa, Carolyn and I passed on the stairs of our zoology building. Hastily I said, "Will you go to the game with me?" She said yes, and so began the most wonderful relationship of my life!

Carolyn spent that Christmas vacation with her parents in Catonsville, MD (near Baltimore), and I with my parents on the farm near

Figure 34. Carolyn Elise Wilson outside the Zoology Building at the University of Iowa in early 1948. Photo by the author.

Figure 35. The honeymooners on top of Bigelow Hill, overlooking the farm that had been owned by Carolyn's grandparents. Photo by the author.

Lamont, IA, where, to earn some money, I trapped furs, mostly mink and muskrat, to supplement my G.I. Bill of Rights stipend.

In the summer of 1948 I drove, alone, in my old Pontiac car to our wedding at the home of Carolyn's grandparents, Elmer and Mina Harris and their daughter Freda. There were no super highways then so it took three days, and, being young, I slept in the car and shaved in filling stations. The family was very gracious, and I think I fitted in well. After the wedding we spent our honeymoon in the family cottage on nearby North Pond (a lake) at Smithfield, ME.

On the way back to Iowa we stopped for lunch at the headwaters of the Sandy River and I caught a very nice brown trout for our lunch. We stayed overnight at a lake resort owned by the family of one of Carolyn's Mt. Holyoke College classmates. The next day we visited Dr. Marvin Meyer at his lakeside lab on Rangeley Lake where he was researching on a tapeworm parasite of landlocked salmon although he was famous for his research on parasitic leeches of fish. Dr. Meyer was gracious as always and we kept in touch until his death many years later. From there to my parents home, Lamont, IA, was uneventful. Carolyn was very good at reading maps and keeping us on course.

While I was graduate assistant to Professor Nolf, Carolyn assisted Professor Robert King in General Zoology. I recall Carolyn helping a deaf student. Where there's a will, there's a way. Professor King was one of our favorite teachers. He was moderately rotund and middle-aged. When the football coach asked him to pass a failing football player in zoology, Dr. King said, "I'll pass him if you'll let me play in Saturday's game!" Another of his favorites was, "If you want to remember people's names, find something unusual about them and relate it to his/her name. For example take Mrs. Commaugh: just think of `stomach' to remember `Commaugh.' But when the student I taught that trick met Mrs. Commaugh, he said, `Hello, Mrs. Kelly!'"

Carolyn spent 51% of my life with me so a large part of this memoir should include her, but I can't remember enough detail. But I will include what I can. She was born in St. Johnsbury, VT, in 1926. Her father was Theodore H. Wilson I. Her mother was Faith Evelyn Harris Wilson of Skowhegan, Maine. Later Carolyn's father earned the PhD degree in Education, served as Head of a Baltimore school for boys and later as President of the University of Baltimore. During this period their home was in nearby Catonsville, Maryland, where Carolyn was No. 1 in that large high school, being voted the most likely to succeed. From this period I have a nice 19 by 25 inch landscape water color she made, and a poem she wrote during World War II which was inspired by a drill after the school had prepared for air raids in 1942 and published in the school paper. It is included here:

AIR-RAID!

Folks do look funny sitting there,
Right on the floor. The walls seem bare
With windows all blocked off by now.
You know, it's interesting how
The kids don't talk so awf'ly loud

During summers Carolyn, her brother Ted and sister Faith usually stayed with their grandparents Elmer and Mina Harris in Skowhegan, Maine. She always loved Maine.

Figure 36. The author, PhD, and his wife, MS, after receiving their diplomas in 1950. Photo by the author.

During her high school days she took violin lessons at Peabody Music Institute in Baltimore and she never forgot how to play her violin. After her death I gave it, a good German violin, to the nearby Shepherdstown College music department where they promised to loan it to new students.

After high school she attended Mt. Holyoke College in Massachusetts, where her major studies were in zoology. She was a member of the well-directed Mt. Holyoke College Choir. She was a very good soprano.

After graduating *summa cum laude* (No. 1) she attended a summer session at the famous Woods Hole Biological Station. In the fall she came to the University of Iowa on a scholarship to study under the late Dr. Emile Witchi, a famous endocrinologist-embryologist, and did her Masters Degree in embryology. I think she received A in everything except an advanced math course. We took two courses together—she bested me in Genetics, but I got top grade in Entomology, primarily because I was more familiar with aquatic insects, including larval forms.

We were married in September, 1948, at her grandparents' home, in Skowhegan, ME, with her father, President of the University of Baltimore and formerly a Presbyterian pastor, officiating. Returning to Iowa, we went to my parents' farm where I could show off my wonderful bride. The first evening there we were loudly shivareed by neighbors and family friends. The loud banging and hollering was followed by all the ice cream and cake we could eat—an interesting farm custom.

In 1950 we took our respective degrees, side by side, at the University of Iowa where the university president declared the PhD the highest degree in course. I have always remembered that!

10. Carolyn's Family and Ancestors

Carolyn's father, Theodore H. Wilson I, was born in 1885 in Middletown, CT, fourth of six children. His father was Edwin H. Wilson who ran a private school in Cambridge, MA. All the male Wilsons before Edwin were farmers, going back to the original immigrant, John Wilson, who came from Northern Ireland to New England in 1729.

Carolyn's father, a very intelligent man, had a remarkable education and career. He took his bachelor's degree at Harvard College in 1907 *magna cum laude*, with majors in Philosophy and Greek. In 1908 he took a masters degree at Harvard College with honors in Philosophy. In 1911 he took a masters degree at Union Theological Seminary, NY, with honors in Religious Education. In 1928 he took a masters degree at the Harvard School of Education. In 1935 he took a Doctor of Education degree at the same school. In 1980 he was awarded an L.H.D, Honorary Doctor of Humane Letters, University of Baltimore.

Ted, as he was known by friends, had an equally amazing career which started in 1908 as a church circuit rider in Alberta, Canada. That was followed by Principal of the Union School of Religion in 1910-11, Pastor of Montclair Congregational Church in 1912-1915, then Pastor of the Congregational Church in Skowhegan, ME during 1912-1915, then Professor of Religion at the Community Church of Olivet, MI, in 1916-1919, then Principal of St. Johnsbury Academy, VT, for the decade 1920-1930 (during this tenure their three children, Faith, Theodore Jr., and Carolyn were born), then Head of Chevy Chase Junior College in Baltimore, MD, in 1930-1932, followed by Head of the National Park Seminary, also in Baltimore in 1932-1935, then head of the Upper School of McDonogh in Baltimore during 1935-1940, then President of the University of Baltimore for 21 years, 1940-1961, and finally professor at Towson State College in Baltimore during 1961-1969. Ted Wilson I was a strong advocate of Junior Colleges and actually served as head of two of them—Chevy Chase and National Park Seminary. At McDonogh School for Boys he helped raise the academic standards in the Upper School so that college acceptance levels became high and remained so.

In addition to all of the above, Ted Wilson Sr. was a strong member of the Rotary Club and Rotary International. He was a charter

Figure 37. Theodore H. Wilson I during his tenure as President of the University of Baltimore. Photo by Dorsey Newspix.

member of the Skowhegan, ME, Rotary in 1923 and President of Baltimore, MD, Rotary in 1948-1949, District Governor of Rotary of MD in 1953-1954, a member of the Board of Directors of Rotary International in 1960-1961, Trustee of The Rotary Foundation in 1965-1957, and finally President of Green Cove Springs Rotary, FL, in 1973-1974.

Carolyn's mother, Faith Evelyn Harris, was born and raised in Skowhegan, ME, and attended Mt. Holyoke College in S. Hadley, MA. Her church was the one where Theodore Wilson I was pastor, and that is

Figure 38. Carolyn's parents at home in 1940. Photo by Carolyn's brother, Ted.

where they met and were married. She encouraged their three children, Faith, Theodore II, and Carolyn to do well academically and to pursue musical skills. Carolyn played the piano and violin and sang, often solo, in church choirs all her life. When her voice started to fail, she bravely played in the church handbell choir. Her siblings also did well musically. Carolyn's mother died much too young of breast cancer.

Carolyn's stepmother, the former Helen Estelle Wright, was head of the nursing school in Dr. Ted Wilson I's University of Baltimore. She and Ted I were married some time after Ted's first wife, Faith, died. She was also a good helpmate. At Penney Farms, FL, she survived Carolyn's dad by several years.

Carolyn's maternal grandfather, Elmer Harris, was in charge of the Skowhegan Jersey Creamery which sold milk, cottage cheese and ice cream. During retirement he created a small American cheddar cheese factory and sold the cheese to grocers. Her maternal grandmother, a very good and loving housewife, died of cancer, but late in life. Their other daughter, Freda, another Mt. Holyoke College graduate and cello player, had been a mathematics professor and Dean of Women at the University of Vermont at Burlington, but retired early to help her parents. In 1952 she married Clyde Allen, a retired woolen mill employee who died in 1953 of lung cancer. Later, in 1966, Freda married a retired carpenter, Clifton Whittemore (Uncle Cliff). Freda was a very kind person who helped take care of her mother and father and her handicapped Aunt Coralie Parker before they died. Freda did well with the cello and she and a local group performed local concerts. Her trusty cello was willed to Carolyn and passed on to my son, Professor G. Lyle Hoffman, who uses it in his Physics of Music class at Lafayette College, Easton, PA. Freda died in 1972.

Carolyn's Sister, Faith Wilson LaVelle, Born 1921.

As far as I know, Faith was A-1 in everything, just like Carolyn and her brother, Ted II. She took her BA in Zoology, *summa cum laude* at Mount Holyoke College in South Hadley, MA, in 1943. Her MA was also in Zoology at Mount Holyoke College, in 1945. She received the PhD in Biology at Johns Hopkins University in Baltimore, MD, with research done at the Department of Embryology of the Carnegie Institution of Washington in Baltimore. She was an administrative assistant in the Department of Zoology of the University of Pennsylvania, 1949-1951; then an Instructor of Anatomy at the University of Pennsylvania Medical School in 1951-1952. Following that year, she was an Instructor of Anatomy at the University of Illinois College of Medicine in Chicago for the years 1952-1968. She became Professor in the Department of Anatomy at Loyola University of Chicago's Stritch School of Medicine for 1968-1987 and Professor Emeritus of Loyola 1987 to present (2010). At Loyola she was Director of Histology and also taught in embryology and neuroanatomy. Her research was in the field of neuronal development and the response of nerve cells to injury. She was the recipient of several teaching awards and was Acting Chairman of Anatomy at Loyola University in 1984-1986. She was National President of Camp Fire Girls in 1978 and 1979 and a Trustee of Mt. Holyoke College in 1960-1965. She and husband Arthur have one daughter, Audrey, born in 1956, now living in Pasadena, CA.

Carolyn's Brother-in-Law, Arthur LaVelle (Husband of Faith), Born 1921.

Arthur took his BS in Biology at the University of Washington in 1946, his MA in Biology at Johns Hopkins University in 1948, and his PhD in Anatomy at the University of Pennsylvania Medical School in 1951. He held a post-doctoral position at same place in 1951-52.

Starting as Instructor in 1952, he rose to Professor in the Department of Anatomy at the University of Illinois College of Medicine in Chicago, and became Professor Emeritus at the same place in 1987. He taught medical, dental and nursing students, primarily in anatomy and neuroanatomy.

His research was in the field of neuronal development and response of nerve cells to injury. He held National Institutes of Health grants for seventeen years and was the recipient of a Guggenheim award for one year. Arthur had sabbaticals at the Department of Embryology at the Carnegie Institution of Washington, and at the Brain Institute of UCLA. There were many published papers and graduate students.

Carolyn's Brother, Theodore (Ted) Wilson II, Born 1922.

Ted took an accelerated college course with no degree at Harvard in 1943, with a major in Biochemistry, and a MD degree and Lt. JG in the U.S. Naval Reserves in 1947 (This was an accelerated program during WWII.). He worked at the U.S. Naval Hospital in Boston, MA, in 1947-1949. Ted was Ship's Medical Officer in 1949-1951 on the U.S.S. Rochester CA-124 with operations in the Atlantic and Pacific Oceans, extending into the early days of the Korean War.

He served again in the U.S. Naval Hospital in Boston during 1951-1954 and then was transferred to the 1st Marine Division in Korea for 1954-1956. Next he was sent to the U.S. Naval Hospital in Oakland, CA, in 1956-1962. His next post was the U.S. Naval Hospital in Bethesda, MD, 1962-1965, as Assistant Chief of Surgery. Next he served on the U.S.S. Repose AH-15 Hospital Ship during 1965-1966, supporting operations in Vietnam as Chief of Surgery. His last position with the Navy was as Chief of Surgery at the U.S. Naval Hospital in Bethesda, MD, for the years 1966-1971. After retiring from the U.S. Navy, Ted served as Chief of Surgery at Union Memorial Hospital in Baltimore, MD, in 1972-1985 and simultaneously as Asso-

Figure 39. Carolyn (left) with her brother Ted and sister Faith in 1981. Photo by the author.

ciate Professor of Surgery at Johns Hopkins School of Medicine in 1974-1985.

Ted's wife Carolyn (Kay) was a physical therapy graduate and a wonderful person. She died of pancreatic cancer in April, 1999.

11. University of North Dakota, 1950–57

In the fall of 1950 we moved to Grand Forks, ND, where I had accepted an Assistant Professorship in the Microbiology Department of the University of North Dakota Medical School. I taught Medical Parasitology and Medical Technology Bacteriology. The funniest incident I recall was when I was walking to work with one of the Med. Tech. students after we had examined rat feces for helminth (worm) eggs. She said, "You mean to tell me that you can tell what parasites are in the rat's intestine just by looking (microscopically) at the rat poop!" My research continued on the parasites of fish, resulting in nine publications while I was there. One highlight of the latter was searching for the final host of one trematode parasite, *Diplostomum baeri eucaliae*, for which I trapped four mink, in season. They were negative for that parasite; however, the pelts netted me $100 which, in those days, was enough to buy a set of china for Carolyn, and the back issues of the *Journal of Parasitology* for me!

Another noteworthy memory of that era for me was the fine technician I had. She, Kathy Piggot, now Mrs. Keith Moodie of New York City, was a young girl in the Caribbean island Antigua during WWII. The U.S. soldiers there took up a collection to help her come to the U.S. for education, which she did. She became a microbiologist. She returned to the Caribbean, married Mr. Moodie, and lived in Jamaica for awhile.

In Grand Forks we were able to continue vegetable gardening which we had started in 1948 when Professor Nolf let us raise a garden at his home in Iowa City, IA. Also in ND, duck hunting almost became an obsession with me. At the beginning, if I aimed at the leader of a flock, I sometimes hit one at the rear. A lady hunter told me to aim at the bird's head, swing with its flight, and fire. It worked, and I filled our rented freezer with ducks and a few Canadian geese. But one time later I aimed at the leader of a flock of geese, and as he went down, spiraling, the flock followed him. I never found him and always felt badly about that incident. Another time, with the family in the car, I grabbed my gun and sneaked up on a flock in a field. I got close enough to be in range, stood up, the flock rose, I aimed, pulled the trigger, and nothing happened. In my haste I had forgotten to load the gun!

During the seven years at Grand Forks, Carolyn gave birth to Pamela, G. Lyle, and Rebecca. She was also active in church where she sang in the choir. She had sung in the Mt. Holyoke College choir earlier and had taken violin and voice lessons at Peabody School of Music in Baltimore. Music was an important part of her life, she did very well at it, and I enjoyed listening to her very much. She had an excellent soprano voice. I miss her.

Graduate Students

I had three good MS students at UND. Jim Hundley did his MS on the development of the anchor parasite *Lernaea* and took another MS in parasitology at, I think, Michigan. He followed his training to a Public Health position in Rochester, IL. Jim was a good student, but the graduate dean made him take a course that wasn't required, and that messed up his schedule. Jim didn't complain and did well. He understood my work and brought a live wild bird to the lab for me to finish a parasite life cycle. One cold day he walked to school in cold North Dakota without ear cover and suffered frostbite on his ears. Francis Ikezaki (deceased) did his thesis on a monogenetic trematode, then went to medical school and became a family physician in his home state of Hawaii. Omer Larson started with me, but transferred to the Zoology Department at the University of Minnesota in Minneapolis, when I moved to West Virginia. His new major professor was the famous parasitologist, Dr. Frank Wallace. Omer (Lars) continued his PhD thesis on larval digenetic fish trematodes. Later he returned to Grand Forks, ND, to join the Biology Department. He published good papers on the trematodes. At about his retirement time, he also became a forensic biologist, identifying larval insects in cadavers to determine the time of death. He was head of the Biology Department for a time.

The first time I met Frank Wallace, Lars' PhD major professor, was in a Paris U.S. Army medical lab in 1944, during WWII, when he was a lab Captain and I a second lieutenant. In his helpful, kindly way, he helped me identify a human tapeworm. And so began a lifelong friendship. At a later time, at a national conference, at dinner he cut Lars' meat because Lars had a broken arm.

Jim Hundley and I learned to duck hunt together. I recall one hunt where we took turns sneaking up on ducks in a marsh. One time it was

Figure 40. Portrait of the author as a duck hunter in 1957. Photo by an unidentified fellow hunter.

his turn while I stayed back. He got within range, flushed several ducks and emptied his gun without scoring! I couldn't help laughing, but was sorry later because I wasn't any better, and it made him feel bad. Later, we both improved.

Figure 41. The University of North Dakota on a typical winter day in 1956. Photo by the author.

North Dakota Friend

Carolyn and I developed a wonderful friendship with Dr. Arthur K. Saiki (deceased), a histopathologist in the medical school. He was originally from Hawaii, as was his friend Francis Ikezaki who became my graduate student. Dr. Saiki's wife had died some years before. When we were en route to Glacier National Park years later, after we had moved to West Virginia, he was a wonderful host on our way through Grand Forks. Some years later he drove out of his way on a trip to a conference to visit us in WV. At the lab in ND, I had a chance to identify human parasites for him. Carolyn and I often thought of him and I still do.

Summer Trips (1950-1957)

During the North Dakota sojourn (1950-1957), our summer vacations took us to Carolyn's Grandparents Harris' and Aunt Freda Harris' home in Skowhegan, ME, including stays at their cottage on nearby North Pond in Smithfield, ME. I usually flyfished for smallmouth

Figure 42. Dr. Saiki (left) and Hal Brody with Lyle and Pam on a visit to our home in Grand Forks in 1955. Photo by the author.

bass in nearby Sandy River. After Grandparents Elmer and Mina died, Aunt Freda was our Maine anchor. She was very kind and helpful.

The car trips from Grand Forks, ND, to Skowhegan, ME, were always interesting. I remember one trip before the children were born. We carried the thirteen foot light canoe on top of the 1943 Chevrolet sedan and carried camp equipment, including my U.S. Army pup tent so we could camp on the way. The interesting, and seemingly best route, took us on US Rte. 2 across Minnesota to Sault Ste. Marie and then to the Canadian National Highway across Ontario to St. Johnsbury, VT, and back on to US Rte. 2 to Skowhegan. On one early trip we stopped at the Algonquin Provincial Park in Ontario and camped

Figure 43. Carolyn beside our car en route to Maine in 1952. Photo by the author.

near Opeongo Lake. It had been a dry season with bears searching for food. We left our food in the car, but in the night a black bear stuck its nose in our tent. Fortunately it smelled no food and left. I had my hand on our camp hatchet and am glad I didn't have to try to use it.

We had a small outboard motor for the canoe and were advised to go to the far end of the lake for large smallmouth bass, but our canoe was so slow we couldn't go that far in the time we had. So no large bass for us!

We also made smaller trips to Iowa to visit my parents Ernest and Viola Hoffman as well as brother Melvin and family in Strawberry Point, IA. Before we left ND we made one very memorable trip to Glacier National Park where we stayed in a small cabin. One very fine day, while Carolyn attended our three children, I went flyfishing in a nearby stream and happened upon a nice pool with many nice cutthroat trout present. I was able to catch enough on floating flies for a fine supper of fresh trout for our family.

12. Fish Parasitology Researcher at the Eastern Fish Disease Laboratory, 1957–1975

In 1957 I accepted the position of fish parasitologist at the Leetown (mail address Kearneysville), WV, lab. In February we sold the ND house at no gain, sent the furniture and other belongings by moving van, packed the two cats and three children (one a baby) in our station wagon, and started in zero degree Fahrenheit weather. The heater in our 1954 Chevrolet station wagon was not working well, so it was a cold trip, but Carolyn, as usual, was very resourceful and kept the children covered with blankets. As we reached West Virginia it was not as cold, but it was snowing. It was not a fun trip, but cats, children, and adults all behaved well. After reaching Charles Town, WV, we stayed in an almost snowbound motel. Dr. Stan F. Snieszko, director of the lab, a kindly person, had reserved the motel for us and also rented a house in Charles Town for us. It turned very cold and Stan accused us of having brought the cold ND weather with us. In about a year we bought a nice one acre lot about a half mile from the lab and hired a small lo-

Figure 44. The Leetown house designed by Carolyn, shortly after it was completed. Photo by the author.

Figure 45. The Eastern Fish Disease Laboratory in 1958. Photo by the author.

cal crew to build our nice house after Carolyn drew the entire house plan on a large piece of brown wrapping paper. She drew the plan in centimeters, but had to redo it in feet and inches because the carpenters were not prepared to work with centimeters. The carpenters did very well, and became good friends, so we had a lovely home for forty-four years.

The lab at Leetown, (post office Kearneysville), WV, was then known as the Eastern Fish Disease Laboratory with a similar one, the Western Fish Disease Laboratory, in Seattle, WA, headed by R. R. Rucker (deceased). Our lab was in a nice stone building with fish tanks on the first floor and lab proper on the second floor. The last group in the lab before our era was headed by Dr. H. S. Davis (deceased), Dr. J. S. Gutsell (deceased), and E.W. Surber (deceased). The fish hatchery was nearby with concrete raceways for trout and earthen ponds for various warm water fish species.

Our main task was to assist National and State Fish Hatcheries in the Eastern half of the U.S. In addition we were encouraged to conduct fish disease research, a good combination. Also, periodically, we held fish disease courses for U.S. and State fisheries personnel and various university researchers. On occasion we had researchers and/or students from foreign countries—Czechoslovakia, Germany, Japan, Peru, Portugal, Scotland, and Ukraine. (Much later, in the 1990s, there

Figure 46. The Fish Hatchery attached to the Eastern Fish Disease Laboratory in 1958. Photo by the author.

were American/Russian conferences at Leetown, headed by Dr. Rocco Cipriano.)

Sometimes long-term scientific associations were made. One example was my contact with Dr. Henrique Mateo of the Universidad Nacional Federico Villarreal in Lima, Peru, who was one of our students about 1970. He, a parasitologist like me, kept in contact with me for many years. When I retired and finished the revision of my book about 1998, I shipped all of my parasitology and related journals to him. In thanks his university named a section of their library after me and made me an honorary professor of his university.

I remember two researchers from Japan who worked in virology for some months. One, Dr. Sano, had bought a new Volkswagen for his use in U.S. and sold it to me at the proper price when he went home. He took great pains in explaining the controls to me and we corresponded about it after he returned to Japan. The other Japanese scientist salvaged a nice eel from an experiment, broiled it in his apartment, and invited me to share. It was good.

Another student, Dr. Fred Meyer, a recent PhD graduate from Iowa State University, was trained as a fish parasitologist and, of course, we

Figure 47. Dr. Malmberg (center) with the author and Dr. Lim at a meeting in Czechoslovakia in 1983. Photo by Plechaty.

became compatriots. He then became the fish parasitologist for the U.S. Fish Farming Experimental Station, Stuttgart, AR, and we published a joint author book on fish parasite treatment and control. That book was so popular that I had to give up my personal copy when I retired.

One PhD student, Joe Hunn, sort of stood out because he had a pet snake for some years. Another PhD student was from Portugal. He published a description of a marine fish parasite after he returned to Portugal.

Another standout visiting researcher was Göran Malmberg, PhD, from Sweden. He was a very helpful friend who became, in my opinion, the leader of his specialty, monogenetic trematodes and, at last account, was still working on them. At an airport in Italy he caught a money changer short-changing me and made her correct it. At an International Parasitology Conference in Warsaw, Poland, he scouted out a very good restaurant that served roast duck, and in Toronto, Canada he found a place that made good shishkabobs. In Munich at a parasitology meeting our group supplied bus service to and from a concert which was a little ways out of town. Göran, another member, and

I missed the bus and we didn't know how to get back to our hotel in Munich. Göran figured it out quickly and got us on a train back to Munich. Such friends were a joy indeed.

Another of our students, John Plumb, later earned his PhD at Auburn University and became the fish bacteriologist there. Tom Wellborn returned to the South to work, but died early. Monty Millard returned to his work in the Midwest and became President of the Fish Culture Section of the American Fisheries Society. He also was a good friend of mine and helped me become one of the first two members of the National Fish Culture Hall of Fame.

Ron Goede returned to Logan and was chief fish diagnostician there for many years. J.T. (Tim) Bowen transferred to our Washington, DC, office where he worked for many years. Roger Dexter returned to the Northeast where he worked for many years. He was our expert on salmon diseases.

Although he was not a student at our lab, Roland Walker, PhD, should be mentioned also. He was a professor at Rensselaer Polytechnic Institute of Troy, N.Y., who became very interested in parasites of bait fish, and we corresponded at great length about his findings. His

Figure 48. The staff of the Eastern Fish Disease Laboratory in 1966. Photo by the author.

extensive findings, including photographs, were forwarded to Dr. Richard Heckmann, Brigham Young University of Provo, UT, about 1987 after Roland died.

The staff and sections of our lab were organized as follows. Dr. S.F. Snieszko was Director. The Bacteriology Section consisted of Dr. Snieszko and Dr. Pete Bullock. Dr. Ken Wolf and his assistant Millicent Quimby were the Virology Section. Histopathology was Ed Dunbar, on his own for several years but later joined by Dr. Marsha Landolt (deceased) who completed her PhD at the Washington Zoo and Dr. Roger Herman who completed his PhD in Ohio. The Parasitology Section consisted of Dr. Glenn Hoffman, Dr. Robert Putz and technician Maria Markiw. Diagnostics for eastern USA were done by Lyle Pettijohn. The library was managed by Florence Wright. More recently the library is managed very well by Vi Catrow.

The secretaries 1960-1975 were Juanita (Nita) Collis, Marianne Strider and Bonnie Knott, all very good workers. Monty Stuckey, our lab assistant, took a degree in photography and became our photographer. Last, but not least, was Winona Owens, who started as a sort-of cleaning lady but learned well, and became a lab technician.

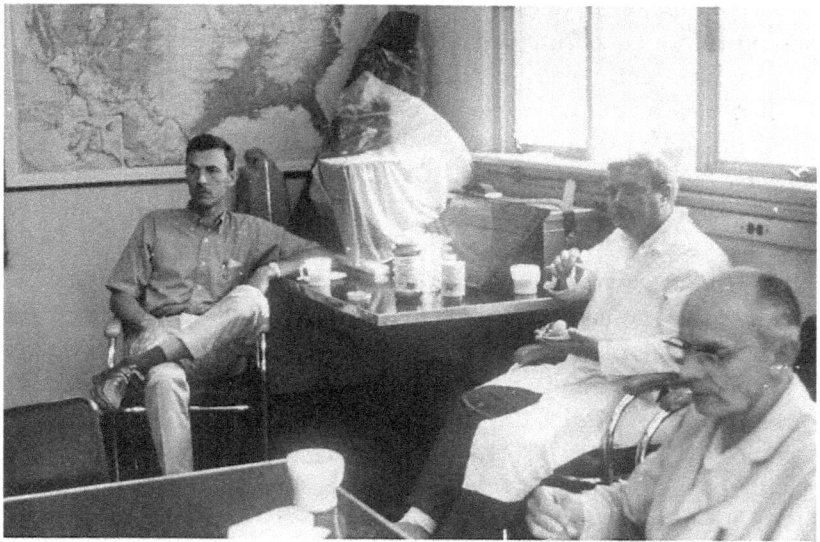

Figure 49. Bob Putz (center) with Monty Stuckey and the author at the Leetown lab in 1970. Photo by Dr. Sano.

Others who progressed in scholastic degree while under Dr. Snieszko were:

Ed Dunbar, who finished his high school degree and bachelors degree followed by training at the nearby Veterans Hospital to become a world expert in fish histopathology. He assisted many of us in the histopathology of our work.

Robert (Bob) Putz, who came to Leetown with an M.S. degree from my alma mater, the University of Iowa, completed his PhD at Fordham University and assisted me in many of my projects. He transferred to our Washington office and some time later became Assistant Director of Fisheries Research and eventually Regional Director of Alaskan U.S. fisheries. More recently he was Director of the Freshwater Institute, Shepherdstown, WV, sponsored by the Conservation Foundation.

Pete Bullock also completed his PhD at Fordham University. He became lab director some time after Dr. Snieszko retired.

Marsha Landolt helped many of us with the histopathology of our work and completed her PhD with a project at the Washington, DC, Zoo. Later she moved to Seattle to become Dean of the Graduate School at the University of Washington. Later, she and her husband, Robert Busch, were killed by a snow avalanche. She was a very capable scientist and it was rumored that she would have probably become a college president.

Roger Herman also worked in histopathology, completing his PhD in Ohio and returning to Leetown to conduct research in histopathology. He became lab director after Pete Bullock retired.

The years at the Leetown Lab, 1957-1974, were very productive for me. I published 16 scientific papers in fish parasitology and the 1967 book, *Parasites of North American Freshwater Fishes* which earned the best book of the year award from the U.S. Fish and Wildlife Service. I gave papers (presentations) at the annual meetings of The American Society of Parasitologists, American Fisheries Society, American Society of Protozoology, The Wildlife Disease Association, and International Verein Limnology (Leningrad). During this period I was helped by many, including my wife Carolyn, who polished up my messy drawings for my 1967 book on fish parasites.

In my section, Parasitology, Robert (Bob) Putz, PhD, was my assistant and Maria Markiw was my technician. Also, for a few months, my

Figure 50. Dr. Marsha Landolt with the author at the University of Washington in 1977. Photo by Carolyn.

son, G. Lyle Hoffman, assisted me with some lab work culminating in the publication of a minor research report. Bob took his PhD while working with me, doing his dissertation on a fish parasite, *Sanguini-*

cola, found locally. The secretaries, Juanita Collis, Marianne Strider, and Bonnie Knott, were very patient with me while typing and re-typing the huge manuscript for the first edition of my book, *Parasites of North American Freshwater Fishes*. I guess they were glad they didn't have to type the second edition! That Christmas my present from them was a book with the entire insides removed and the cover was titled, *Glenn's next book!*

The most exciting project while I worked at the Leetown Lab was verifying Dr. Snieszko's presumptive diagnosis of whirling disease of trout caused by a protozoan parasite, *Myxobolus cerebralis*, known only in Europe before that time. On the basis of the disease signs, trout whirling at time of feeding and black tails, Dr. Snieszko found the cause in a German fish disease book. I had only to scrape the spores out of the equilibrium organ to verify it. The spores were identical to those pictures in the German book. The parasite had come to the U.S. in frozen rainbow trout from Denmark. Ed Dunbar (deceased) helped me demonstrate the effectiveness of using ultraviolet light to disinfect water contaminated with *Myxobolus cerebralis*. He also prepared the histological slides. He was a gem.

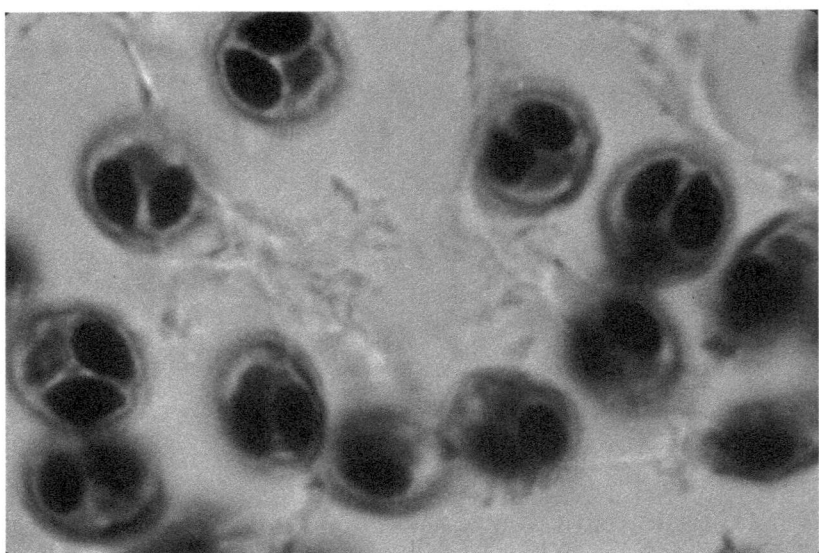

Figure 51. *Myxobolus cerebralis* spores, Giemsa stain. Photomicrograph by the author.

Whirling Disease Spread Widely

About the time of our identification of the cause of this imported parasite, it spread rapidly because we didn't know its life cycle. One early transfer was to northern Michigan where it spread from an initial transfer spot to another site, resulting in one establishment suing another. I was called to be an expert witness on two occasions. I don't remember the details, but the side for which I testified won both times. The whirling disease continued to spread and, whereas we believed it not to be dangerous to wild trout stocks, it became of concern in wild trout populations in western USA. Eventually a coalition of western concerned parties was formed and research on whirling disease continued. During this period, in 1978, R.E.L. Taylor and M. Lott of the University of Nevada demonstrated the spread of whirling disease spores by birds, a very important detail.

More on the Eastern Fish Disease Lab

As stated above, our lab, in addition to assisting U.S. Govt. fish hatcheries, also assisted state and private fish hatcheries, and in some cases they provided information for us, a good exchange. I recall particularly Doug Mitchum at the Wyoming fish disease lab. He and I exchanged fish parasite information for many years. He was always pleasant and helpful.

The fourteen years at the Leetown Lab were very good for me. Dr. Snieszko was an excellent director who understood what researchers needed to do and he helped them do it. For example, in 1966 I told him that I would like to write a book on fish parasites. He said, "If you want to become famous, write a book; you may have one year off of lab work to do it." So, the first edition of *Parasites of North American Freshwater Fishes* was written and published in 1967. While I was writing the book, Dr. Bob Putz kindly handled laboratory matters with Maria Markiw assisting.

During those fourteen years, others at the lab also helped me. Ed Dunbar, mentioned above, made the histological preparations for several of my projects, and later Marsha Landolt and Roger Herman did likewise. The librarian, Florence Wright, helped find many publications for my use, and the secretaries mentioned earlier did very well

Figure 52. Carolyn and Dr. Oleg Bauer in Leningrad in 1971. Photo by the author.

in raising my writing from a scribble to neat typed pages. The late Professor Dr. Oleg Bauer of the Zoological Institute of Leningrad (now St. Petersburg) supplied me with helpful Russian fish parasite literature, most notably *The Fish Parasites of the USSR* which I was able to get translated through a U.S. Government translation program. They provided me with twenty copies that I distributed to American colleagues. The Russian book helped me greatly in shaping my book on American fish parasites. Dr. Bauer also invited me to participate in an international fish parasite conference in Leningrad which was the beginning of a wonderful friendship that continued for the rest of his life. He died in year 2003. One of the reasons for our wonderful collaboration was his command of several languages, including English! I owe him very much for my progress in fish parasitology.

An interesting little side note about the identification and naming of the fish parasites: Our classification system of naming animals, including parasites (Phylum, Class, Order, Family, Genus, Species), began with the great Swedish naturalist, Carl von Linné, commonly known as Linnaeus. My good dentist in Charles Town, WV, Dr. L.R. Jackson, was so impressed with the classification system that he, presumably with his wife's concurrence, named their daughter Linné!

Figure 53. Dr. Jiri Lom with the Hoffman family on a visit to Leetown in 1966. Photo by the author.

Another wonderful collaboration for many years was with Dr. Jiri Lom of the Institute of Parasitology, Czech Academy of Science, České Budějovice, Czechoslovakia,. He is probably the preeminent fish protozoologist of the world, and he helped me with fish trichodinids and sporozoa for many years. He was able to visit our lab, and I his, and he also organized the first International Fish Parasite Conference which became well established. He and his colleague, Dr. I. Dykova, published *The Protozoan Parasites of Fishes* in 1992. Dr. Lom was tall and

Figure 54. Dr. Stan Snieszko after his retirement. Photo by Pete Bullock.

I short, so we might have been called "Mutt and Jeff!" I will always remember his expertise and kindness.

In addition to the two fish disease research labs there were smaller USFWS diagnostic labs at other locations, N.E. U.S., S.E. U.S., Colorado, Utah, Oregon, and Washington. The one at our lab was a one-person section served by Lyle Pettijohn. He received incoming samples and also traveled to New England, helping wherever he could. It was a good relationship because we continually exchanged diagnostic information. I always remember Lyle asking me to check something microscopically, and if I needed samples he obliged graciously. If Lyle found something unusual he would bring it to the attention of the most logical researcher. Also, if any of the other diagnosticians in our Eastern Half of U.S. found important items they would bring it to our attention. It was a good system.

How things change. After Dr. Snieszko retired we had a new director who tried to change minor items. In my case, for example, I had saved baby food bottles for many years to use to hold snails to search for trematode cercariae liberated by live snails in the water in the jar. The new director disposed of them, explaining that only professional scientific equipment was to be used in the lab now. Not only would the baby food bottles save money, they were just right for the purpose. Anyway, you win some, and you lose some!

13. Scientific Collaboration, 1960-1975

Our chief, Dr. S. F. Snieszko, favored collaboration with scientists in our and other labs and I greatly benefited from it. Briefly listed here are my major collaborations:

1. The former Leetown chief, Dr. H. S. Davis and his colleague Gene Surber helped on a *Sanguinicola* (blood fluke) project (1961).
2. Ed Dunbar, our histopathologist, assisted me on four projects—a larval cestode in brook trout, with Art Bradford (deceased); the first scientific report of whirling disease (*Myxobolus cerebralis*) of salmonids in the U.S. (1962); studies on *Neogatea*, a larval trematode (1963); and with Dr. Ken Wolf and Dr. Zwillenberger (Switzerland), the virus *Epitheliocystes* of fish (1969).
3. Dr. Carl Sindermann, National Oceanic Administration, helped with a description of common parasites of fishes (1962).
4. Dr. Robert Putz, my assistant coauthored nine reports with me: *Gyrodactylus*, *Pleistophora* (microsporidia), *Uvulifer* (black spot), *Myxosoma cartilaginis* of bluegills, *Urocleidus* (skin fluke of fish), whirling disease (earliest susceptible age of trout, effect of various agents on the whirling disease spore and effect of freezing and aging on whirling disease spores) (1963-1971).
5. Dr. S.F. Snieszko and I co-authored two papers on control of fish diseases (1963 and 1970).
6. Dr. Jiri Lom, Institute of Parasitology, Academy of Sciences, České Budějovice, and I co-authored two papers, on Trichodinids (1964) and *Myxobolus cerebralis* (whirling disease) (1971). Dr. Lom was probably the best fish protozoologist in the world, and a very kind person.
7. Dr. G. W. Prescott, Michigan State University, assisted on a parasitic alga in bluegills (1965).
8. J. A. Hutcheson, a former student of ours, assisted on *Posthodiplostomum minimum* (white grub) in striped bass (1970).
9. Professor Dr. O. N. Bauer, Zoology, Academy of Sciences, St. Petersburg, Russia, co-authored a joint paper with me on fish parasites

in water reservoirs (1971); but more important, he helped provide access to Russian fish parasite publications, including the Russian Key to USSR Fish Parasites which I had translated.

10. Dr. G. Lyle Hoffman Jr., my son, when a high school senior, helped with "Effect of Chemicals on Whirling Disease Spores" (1972).
11. Dr. E. V. Raikova, Institute of Cytology, Academy of Sciences, St. Petersburg, Russia and W. G. Yoder, Michigan Dept. of Natural Resources, Grayling, MI, assisted with the report and description of *Polypodium* species (Coelenterata) in North Amererican sturgeon, a first (1974).
12. Dr. Tom Sawyer (National Oceanic Administration), John Huath (Michigan Deptartment of Natural Resources), and John Conrad assisted in the description of a parasitic amoeba in fish and later named it after me (1975).
13. Dr. Marsha Landolt (deceased), when working at Leetown, I and four others described an unusual fish disease caused by a *Tetrahymena* species (1975).
14. In 1975 I assisted on a report on *Henneguya* of catfish with Joseph McCraren, Marsha Landolt and F. P. Meyer (all of the USFWS). McCraren was the first author.
15. Dr. Fred Meyer, USFWS, Stuttgart, AR, assisted me on the book, *Parasites of Freshwater Fishes: a Review of their Control and Treatment* (T.F.H. Publications, 1974). Dr. Meyer was a very good colleague.
16. Another long collaboration was with Doug L. Mitchum who published *Parasites of Fishes of Wyoming* in 1995.

SUMMARY of these collaborations: It was a very real pleasure to work with these very fine people.

14. The Hoffman Offspring

Daughter Pamela graduated from high school as salutatorian (2nd) in 1969, from Mt. Holyoke College in 1973 and from the Medical School of the University of Virginia in 1978, later becoming Chief of Geriatrics at St. Vincents Hospital in Bridgeport, CT. She has continued to provide medical services to the disadvantaged people of Bridgeport through the Outreach Clinic at St. Vincents ever since.

Son G. Lyle graduated valedictorian (No. 1) from high school in 1970, earned an AB from Dartmouth College in 1974 and a PhD in Physics from Cornell University in 1983. In 1978, Lyle married Kathy Zadorozny, a Master of Fine Arts student at Cornell University. After Lyle took the PhD in Physics in 1983, they moved to Easton, PA, where Lyle started as Assistant Professor in the Physics Department at Lafayette College. His astrophysics research has continued at the world's largest radio-telescope in Arecibo, Puerto Rico. Now (2010) he is a Professor of Physics, having been head of the department at Lafayette College for a six-year term.

Daughter Rebecca (Becky) graduated as salutatorian from high school in 1973 and from Wesleyan University of Middletown, CT, with a Chemistry Major in 1977. She married Todd Craig, PhD in Chemistry, the same year. They moved to Raleigh, NC, where Todd and Becky worked at Union Carbide Agriculture Products in the Research Triangle 1981-1986. They transferred to near Charleston, WV, for the years 1986-1993. Then Todd's Union Carbide Division was bought by Arco Chemical Co. Later Todd transferred to Dennis Chemical Co. in St. Louis, MO. Dennis Chemical Co. was bought by Geon which merged with MA Hanna Co. to become Bay One Polyurethane Systems. They work with polyurethane products. Todd's present title (2008) is Product Manager.

Becky, who had worked at pharmaceutical labs in Connecticut and North Carolina, quit working to start the family additions in 1987. Robin was born in 1987, Heather in 1989, and Kelly in 1991.

When Todd and Becky worked in Raleigh, NC, for vacations they rented a beach house on Emerald Isle, N.C., and they have done so ever since, graciously inviting members of both families to share their

Figure 55. Family portrait dated 1974. The children, from left to right, are Lyle, Pamela, Rebecca and James. Photo by the author.

vacations with them. In 2004-2005 they had a German exchange student, Laura Kunz from Bonn, Germany, who, among other things, was on the track, gymnastics and wrestling teams. She is back in Bonn, Germany, now (2006) and teaches gymnastics, etc.

Son James was born in 1960, started school in WV and continued high school in Stuttgart, AR. When he was a junior he was accepted at Dartmouth College; however, he opted for a year as an exchange student in Bederkesa, Germany, where he did well, including playing the organ on Sundays in a nearby church. Upon return he attended Dartmouth College, Hanover, New Hampshire, where he graduated with Honors in Comparative Literature in 1982. Two months later he was killed on the Hanover/Norwich bridge by a drunken driver.

James was an early author. When he was seventeen, a poem of his was published in *Bardic Echoes*, Vol. 18, No. 1, p. 15 (Jan. 1977).

After son James was old enough to fend for himself, Carolyn took a biology teaching job at nearby Shepherd College (now Shepherd University), then was asked to teach basic science courses in the Nursing

Program. She considered that if she were to teach nurses she should add a degree in registered nursing to her B.S. and M.S. in Embryology, and she did.

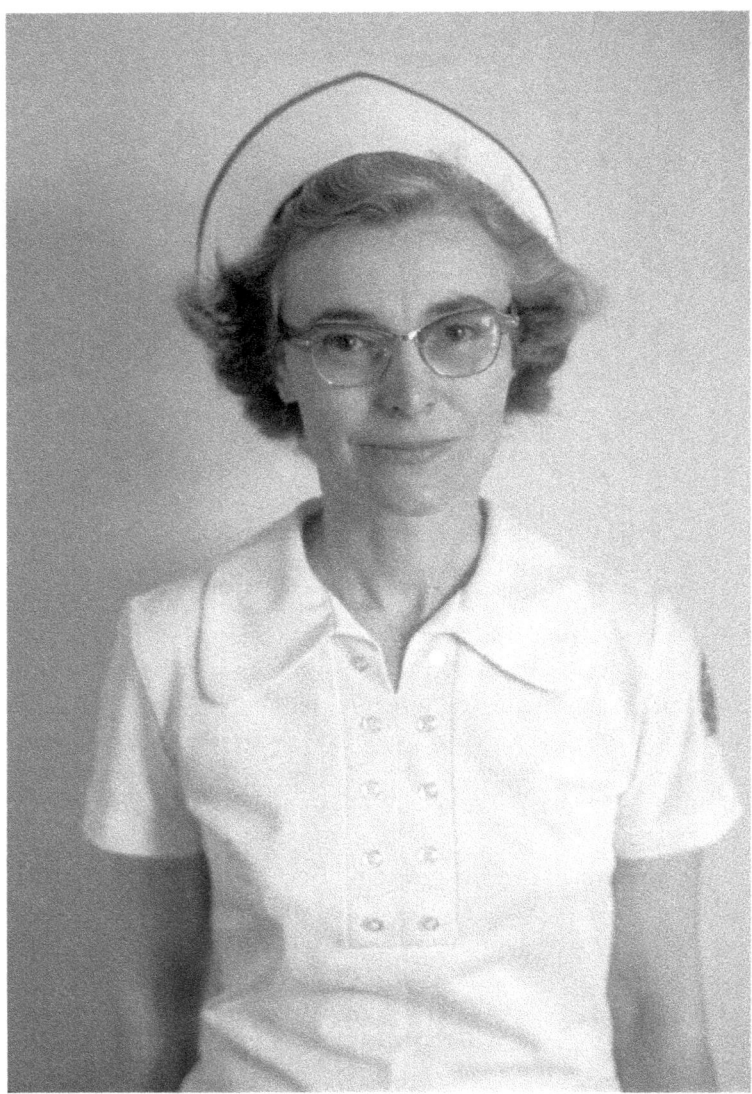

Figure 56. Carolyn, RN, in 1975. Photo by the author.

15. U.S. Fish Culture Station, Stuttgart, AR

About the time Carolyn was studying for her RN degree, in 1974, I transferred to the U.S. Fish Culture Station (now Harry Dupree National Aquaculture Research Center) in Stuttgart, AR, leaving Carolyn to complete the last year of her nursing work at Shepherd College and Rebecca (second daughter) to finish high school in West Virginia. Daughter Pamela was in medical school in Charlottesville, VA and son Lyle was in Dartmouth College, Hanover, NH. This was before James was killed by a drunk driver, so he and I moved to Arkansas where he did the cooking and I washed the dishes. At the tender age of 15, James did well at the cooking and very well at school. Carolyn and Becky moved to Stuttgart to join us after Becky finished high school the next year. At the lab, unpacking the equipment, specimens, library, etc., took a lot of time. There ought to be a better way.

In Arkansas James took organ lessons in Little Rock where he did well. Later, when he was an exchange student near Hamburg, Germany, he became a local church organist at the tender age of 16. He was a quick learner.

The staff at the Stuttgart station consisted of Dr. Harry Dupree (deceased), Director; Don Greenland, Assistant Director; Mayo Martin, Extension Biologist who helped fish farmers with trouble shooting as well as advising about almost anything in fish culture; Drew Mitchell, my assistant, who was well trained by Drs. John Plumb and Bill Rogers at Auburn University, Auburn, AL; Brenda Moore (formerly Rodgers) who did very well as my technician; Vickie Houghton and Patsy Mitchell, secretaries; Joyce Cooper, Brenda's mother, librarian; Dr. Billy Griffin and Dr. Tom Brandt, researchers; Dewey Tacket (deceased) , aquaculture chemist; Allen Thomas (deceased) and Jim Ellis, fisheries biologists; Tommy Mitchell, Ray Carter, Dale Jameson, Les Gill and Kurt Colburn (deceased) , technicians and maintenance workers.

After I retired in 1985, Drew Mitchell continued there as a researcher and has done well. Two review research publications by him include a 91 page review of fish health and diseases in the United States from 1906 to 1969, and the "Public Sector Role in the Establishment of Grass Carp in the United States." Private workers as well as

Figure 57. Drew Mitchell (left) and Brenda Moore with the author in his lab at Stuttgart in 1977. Photo by an unidentified colleague.

government workers were responsible for the accidental spread of grass carp, which in some waters are undesirable.

I was to be at the Arkansas lab for ten years, so transferred my fish parasite researches from mostly trout to mostly pond fishes: catfish, buffalo fish, European grass carp, golden shiners, European grass carp and golden shiners. Resulting publications included research on microsporidian and myxosporean protozoa, ciliated protozoa, trematodes, tapeworms, exotic parasites including *Polypodium*, a coelenterate in sturgeon eggs, fish disease diagnosis and control as well as fish parasites and environment, food-borne parasites, transfer of fish parasites and use of chemicals for control. My assistant, Drew Mitchell, in addition to assisting in much of the above, tested several chemicals for parasite control, including Dimilin, the anti-Gypsy Moth chemical, which we found was effective in controlling the very damaging crustacean, *Lernaea cyprinacea*.

I was chairman of the Translation Committee of the American Society of Parasitologists and was able to get several foreign books translated and published, including the Russian fish parasite book mentioned earlier and the two-volume *Fischkrankheiten* (Fish Diseases) by Schäperclaus, 1986. Several copies were distributed to colleagues. I was on the editorial board of four journals during this period.

Travel to fish disease labs and research presentations took me to Australia, Canada, Czechoslovakia, Denmark, England, France, Germany, Italy, Mexico, Russia, Scotland, Sweden, and most states in the U.S. This was very helpful to me and there was a bonus: my wife, Carolyn, was able to accompany me several times, paying her own expenses, of course. She enjoyed travel very much. Dr. Dupree was very supportive of my travel as was Dr. Snieszko, WV, before him.

During the Arkansas tenure, at one point, because it was a Government lab, our secretary, Vickie, posted this on the bulletin board.

OVERHEARD IN A GOVERNMENT OFFICE.

Secretary:
  I know that you believe that you understand what you think I said, but I am not sure you realize that what you heard is not what I meant.
  The above is just to normalize your thinking for today. It is really the way to think these days.

Our director, Dr. Harry Dupree, at the Arkansas Station, believed, instead of belonging to the USFWS, our station ought to be in the U.S. Agriculture Service. He apparently presented evidence of our relationship to agricultural aspects. After Dr. Dupree died, some years later, the station was transferred to the U.S. Department of Agriculture and named after Dr. Dupree.

During my work at the West Virginia Lab and the Arkansas Station, I was blessed to have visiting scientists, some in collaborative research, from California, Chile, Costa Rica, Czechoslovakia, France, Germany, Iowa, Italy, Pennsylvania, Peru, Poland, Portugal, Russia, Sweden and Wyoming. These were all wonderful co-workers and friends.

During this period I became acquainted with Doug Mitchum's work for Wyoming Fish and Game. Doug worked on their fish parasites for many years and we compared notes regarding the many parasites he identified. I can't recall if he published his results. He became a good friend.

Evolution of Parasites

During all the studies on species *et cetera* we often wonder what particular species might have evolved from. The simplest, and perhaps accurate, explanation can be found in the following quote from *Readers Digest*, May, 1996, p. 110:
"The path of evolution is merely a walk through the realm of possibilities."

About 1980, while we were still in Stuttgart, AR, Carolyn attended a nurses' meeting in a conference facility in Waterville Valley, NH, and I accompanied her, carrying fly rod and creel on the airplane. The conference building was very adequate and we were housed in a complex somewhat like the youth hostels of Europe, entirely adequate. The dining room was also very adequate.

While Carolyn was at the conference sessions, I was free to search for trout streams and found the Mad River nearby. It was summer and I had fun catching nice brook trout on dry (surface) flies, but not many at one outing. The kitchen crew at the conference center kindly saved two or three daily catches for me and later cooked a very nice trout dinner for Carolyn and me. Carolyn said they had a good conference and I had some good trout fishing, a good combination.

Awards

I received some honors over the course of my career. I received the Distinguished Service Award from the Wildlife Disease Association, 1976; the S.F. Snieszko Award, Fish Health Section, American Fisheries Society, 1982; the Meritorious Service Award U.S. Department of Interior, 1983; and the Distinguished Service Award, Fish Culture Section American Fisheries Society, 1985. I was inducted into the National Fish Culture Hall of Fame, Spearfish, SD, in 1985. For my 1967 book, *Parasites of North American Freshwater Fishes*, I received two outstanding publication awards, one from the Wildlife Society in 1967 and one from the U.S. Fish and Wildlife Service in 1968. Nine new fish parasites were named after me, an honor bestowed individually by my colleagues. In year 2000 I was named Honorary Professor by the Universidad Nacional Federico Villarreal in Lima, Peru. Earlier, in 1987, I was given a sheepskin "diploma" "Por su labor transcendental en el estudio de la parasitologia de pesces. Alere Flammam Veritatis" from the Faculty of Biological Sciences of University Autonoma, Nuevo Leon, Monterrey, Mexico where I gave several lectures on fish parasites. It was there after my lectures that a young lady student kissed me for my presentations, the only kiss I ever got for talking about my fish parasites!

16. Those Who Helped Me

During my training and forty year career, many people helped me in many ways. Following is my attempt to document those valuable events. Some have been mentioned in earlier chapters, but bear repeating:

1936-1937.

School Superintendent Paul Young, of Lamont, IA, moved to a large high school in Davenport, IA, and became Dean of Boys or something comparable. He kindly invited me, age 18, to come and live with his family, start college at St. Ambrose College, and help with the high school wrestling program. I had been a Lamont High School wrestling champion at 125 lbs. At St. Ambrose my grades were not of the best, but one highlight was taking Music Appreciation which did what it was supposed to do even though I can't sing or play any musical instrument. I will always "love" good music, particularly classical. During that year I worked part time, in order, as a movie usher, bakery hard roll deliverer, and clerk in Jim Creighton's sporting goods store.

1937-1938.

Because of lack of money, I moved back to Lamont, IA, trapped furs that fall, worked as a farm hand, and also at the State Trout Hatchery where the Superintendent Bob Cooper helped me. Of course my parents, Ernest and Viola Hoffman, always helped me. They taught me to be a good person and the importance of hard work.

1939-1942.

I transferred to the University of Iowa, in Iowa City, and became a zoology major. There my grades improved steadily and I had part-time employment in housekeeping at the University Hospital. The chief of housekeeping was a kind lady named Winona Ballantyne who, because of my allergies to linen dust, transferred me from folding linens to a very good part-time job washing the door windows throughout the

THOSE WHO HELPED ME

hospital. This was a big help because she kindly allowed me to do the three hour per day job as fast as I could which became much less than three hours. I was known as the fellow who had that good board job!

1941-1991.

During my senior year, while taking General Parasitology under the late Dr. L. Owen Nolf, I discovered that I wanted to become a fish parasitologist. Prof. Nolf recognized my desires and kindly pointed me in the direction of fish parasite research. From that time on until his death in 1991, he remained my friend and teacher. For his help I owe him great gratitude and dedicated my 1967 book, *Parasites of North American Freshwater Fishes*, to him. The 1999 revision was also dedicated to him and to my beloved late wife, Carolyn. The latter helped me in many ways, including the revision of the book and the 47 years of our happy married life when we never had an argument!

1946-1950.

During studies and research for my PhD program, I was not doing graduate quality work in Comparative Anatomy. Professor Jerry Kollros (deceased), the Anatomy teacher, recognized my problem, and, knowing of my extreme interest in fish and their parasites, found a way to help me. He assigned a book on fish anatomy to me and shaped the exam around fish anatomy, and I did well. He also helped my wife Carolyn who had a scholarship to do her MS under world famous Professor Emile Witsche who received an unanticipated call to teach overseas that year. Dr. Kollros kindly accepted Carolyn as his student, and she did her dissertation on the development of a peculiar fat body in chickens. I will not forget Dr. Kollros' kindness.

1950.

After I completed the PhD in Zoology in 1950, Prof. Nolf's helpful recommendation gained me a good job teaching medical parasitology and researching in fish parasitology at the Medical School of the University of North Dakota in Grand Forks, ND.

1957.

Recommendations from Prof. Nolf and others helped me to become the fish parasitologist at the Eastern Fish Disease Laboratory in Leetown, WV.

1957-1985.

Prof. Dr. Oleg Bauer of the Zoological Institute of the Leningrad (now St. Petersburg) Academy of Sciences helped me by sending me important books in Russian fish parasitology. I was able to get some of these translated into English by the U.S. Science Translation Program. One, in particular, the *Key to Fish Parasites of the U.S.S.R.*, has been very useful worldwide.

1966-1967.

During my time at Leetown, our lab chief, Dr. S. F. Snieszko, helped me by allowing me a year off from other duties to write *The Parasites of North American Freshwater Fishes* (1967). Being accomplished in many aspects, my late beloved wife Carolyn assisted in the many illustrations in the book. My assistant, Dr. R. Putz, helped me during this period especially by taking over my other duties while I was working on the book. Carolyn also typed the 1999 revision of *Parasites of North American Freshwater Fishes* and, recalling her high school Latin, corrected errors in the Latin fish parasite names. Dr. Putz' secretary, Barbara Rinehart, typed the final corrected copy. I also thank all the other secretaries who typed my longhand manuscripts all those years.

During my thirty nine year career, many people helped me with publications and research. Some of those, in alphabetical order, are: Oleg Bauer (deceased), St. Petersurg, Russia; Art Bradford (deceased), Benner Springs, PA; Jimmy Camper (deceased), Heber Springs, AR; Ed Dunbar (deceased), Leetown, WV; Bernard Fried, Easton, PA; J. E. Harvey, Linesville Fish Cultural Station, PA; S. L. Kazubski, Warsaw, Poland; Don Klemm, Cincinnati, OH; Jiri Lom, Prague, Czechoslovakia; Medical students 1950-1957, University of North Dakota, Grand Forks, ND; F. P. Meyer, LaCross, WI; M. C. Meyer (deceased), Orono, ME; A. J. Mitchell, Stuttgart, AR; D. L. Mitchum, Laramie, WY; L.

Owen Nolf (deceased), Iowa City, IA; J. O'Grodnick, Benner Springs, PA; R. Putz, Shepherdstown, WV; E. V. Raikova, St. Petersburg, Russia; G. Schubert (deceased), Stuttgart, Germany; C. E. Smith, Bozeman, MT; S. F. Snieszko (deceased), Leetown, WV; Roland Walker (deceased), Troy, NY; and W. G. Yoder, Grayling, MI.

17. Foreign Trips to Conferences and Visitations to Fish Parasite Labs

Puerto Rico 1966

At the annual meeting of the American Society of Parasitologists in San Juan in 1966, I gave a paper on fish parasites, but one of the highlights was the wonderful sweet pineapple stuffed with delicious salad. I roomed with my colleague, Dr. Fred Meyer, then from our sister lab at Stuttgart, AR, who had been one of our earlier fish disease trainees at Leetown. Another highlight was a side trip to the tropical rain forest El Yunque (the anvil) in Eastern Puerto Rico. Four of us rented a Volkswagon car and drove there. It was a beautiful place with a nice stream that reminded me of West Virginia bass streams. Always interested in snails, partly because they are host to many fish trematodes, I noted small snails on the shells of larger live snails. Because of the low calcium, the smaller ones were eating the shells of older ones, possibly very unusual. Also there were larger land snails on tree trunks, so I saved some to take home. I forgot to remove them from the VW. I went back a little later to retrieve them, but was told they were bad and so had been destroyed, presumably because of fear of schistosomiasis which could be in water snails, but not land snails.

Europe 1971

Professor Dr. Oleg Bauer (deceased) of the Zoological Institute of the Academy of Sciences, Leningrad (now St. Petersburg) invited me to give a paper on fish parasites at the International Congress of Limnology in Leningrad in 1971. Interestingly and adequately, the conference was held in the Bull Castle. On the way in to Russia, customs officials confiscated my wife's banana and questioned her as to how can one condense a book such as a *Readers Digest* condensed book.

In Leningrad, Carolyn and I visited Prof. Bauer's lab which was much like similar labs in the U.S. However, the museum in the Zoological Institute was very striking, with a mounted wooly mammoth which had been discovered in the perma-frost by dogs, but salvaged in-

Figure 58. A vista in the El Yunque rain forest, PR, in 1966. Photo by the author.

tact! Also there was the mounted horse of Peter the Great. Peter was a large man, but the horse was rather small. The desk of Peter, however, was very tall.

Figure 59. From left to right: Drs. Chubb, Hoffman, Reichenbach-Klinke, Bauer, Furtado and Vik at the Zoological Institute in Leningrad in 1971. Photo by Carolyn.

Carolyn, with no previous knowledge of Russian, learned the Russian alphabet quickly, and was able to use the city transport system and manage the streets for some sightseeing on her own. However, I did accompany her to the world famous Hermitage Museum where I marveled at the large column of malachite green made by cementing small pieces together because malachite green is found only in small pieces. Solutions of malachite green have been used successfully in fish parasite control. Much later, Professor Bauer's multilingual son became the foreign representative for the famous museum, the Hermitage!

My boss, Dr. S.F. Snieszko (deceased) believed that international contact with colleagues was desirable so he approved side trips to labs in Sweden, Denmark, Germany, France, Italy, and England. Later, I was classified "world-class fish parasitologist."

In Stockholm, Sweden, we visited Prof. Göran Malmberg in his lab at the University of Stockholm and his home in Taby. Göran is one of the world's authorities on monogenetic trematodes of fish. We also enjoyed his nice home in the country. He had visited our lab in WV.

In Denmark, we visited a trout hatchery which was operated by a well-informed veterinarian, Dr. N.O. Christensen, who showed us how they controlled fish diseases such as whirling disease (Myxobolus cerebralis). We also enjoyed Danish pastries while there.

In France we visited the lab of Dr. Pierre de Kinkelin at the Station de Virologie and D'Immunologie Moleculaires in the outskirts of Paris where I consulted with him about the pathogenic fish trematode he was working on. He also took us to the home of my WWII landlady Mme. Louise Coutant where she provided a family reunion banquet. She died two years later.

While in Paris we also visited the National Museum where I had taken a picture of the statues, Venus de Milo and Winged Victory, during WWII. On the return on foot, we approached Place de la Concorde (or similar) when it had been raining lightly. There was a beautiful rainbow over the Place!

In Germany we visited Professor Dr. Heinz Reichenbach-Klinke (deceased) of the University of Munich, a world's expert on tropical fish parasites. I presented a seminar there and later Dr. Reichenbach-Klinke took us to a trout hatchery to show us how they controlled the whirling disease of trout. At his home he showed us his fossil collection. Interestingly, the U.S. Army lab where I worked during WWII was only a short distance from his home.

After our visit to Germany we drove through Switzerland to Italy to visit the government lab where Dr. Pietro Ghittino (deceased) worked. Pietro and family had been on a family outing the day before so we were treated to a delicious frog leg dinner. The next day Pietro took us north toward Mount Blanc to visit a trout hatchery where they cultivated the most highly colored rainbow trout I've ever seen. On the way, on a four lane super highway, a car pulled alongside and the driver shook his fist at Pietro. I said, what did he do that for? He said, "Oh, Glenn, those Italians are all like that!" But Ghittino lives on—his son, Claudio, now works at his dad's lab.

In Czechoslovakia we were to visit Professor Dr. Jiri Lom at his fish parasite lab in Prague. However, he had to be away, so his colleague Professor Dr. Frank Moravec was our gracious host. Dr. Lom is a world leader in fish protozoa and Dr. Moravec is a world leader in fish nem-

Figure 60. Val d'Aosta Regional Icthyological Establishment, Italy, in 1971. Photo by the author.

atodes. Dr. Moravec gave us a nice tour of the lab and also provided transportation, including a visit to the famous Karlstein Gothic Cas-

Figure 61. Dr. Ghittino outside his lab in Italy in 1971. Photo by the author.

tle, which was founded by the Bohemian King and Roman Emperor Charles IV for the safeguarding of the coronation jewels. It was reconstructed in the 19th century into a Neo-Gothic luxurious residence for the successor to the former Austro-Hungarian throne, Ferdinand d'Este. This must be one of the oldest existing castles!

This 1971 European trip was very rewarding in many ways. It was particularly good to have one-on-one visits with colleagues I had known only by publications and correspondence.

Europe 1974

At the 1974 International Congress of Parasitology in Munich, Germany, I gave a report on Whirling Disease (*Myxobolus cerebralis*) in the United States. I called it my Swan Song because I would soon be moving to our Stuttgart, AR, lab. Professor Vik of Norway said, "No, this isn't your swan song." I retired eleven years later after many research projects. Dr. Vik was a good friend.

On our day off at the Conference, Prof. Reichenbach-Klinke had arranged a wonderful bus trip to Kochel See, Walchen See, Garmisch Portenkirchen, Linderhof Castle, Oberammergau, Ettal Monastery, and back to Munich by autobahn. I felt honored that my seat companion was the famous eighty-six year-old Horace Stunkard, who was a good companion.

We saw the Passion Play House at Oberammergau, but didn't have time to go in. The highlight of the trip was Linderhof Castle. King Ludwig II had this magnificent castle built only eight years before his drowning. He may have been crazy, a sort of recluse bachelor king. He spent as much Bavarian money as he could on **gold** furnishings and expensive works of art to embellish the rather small castle, a sort-of-miniature Chateau de Versailles. One of our group, Dr. Carl Herman, remarked that a certain table, inlaid with jewels, etc., carries the name of the man who bought it instead of the man who created it! The king had been to a cave with a little lake in it on the Italian coast, which he enjoyed so much he had his subjects dig one for him behind the castle. He had a small boat on the little lake inside and rowed around while his small orchestra played for him! There was a small natural lake nearby where the king, alone, rowed around frequently. He drowned there, but no one knows how it happened.

One side trip on this conference trip was to visit Dr. Nikola (Mike) Fijan of the Veterinary Faculty at Zagreb, Croatia (then part of Yugoslavia) who had worked at the University of Arkansas Research Project earlier where we had met.

Figure 62. Dr. Firjan in his Zagreb office in 1974. Photo by the author.

After arriving in Zagreb, Mike gave me some history of Zagreb. Part of what used to be known as Yugoslavia uses Latin letters, part uses Cyrillic, like Russian. The spoken language reminded me of a cross between Czech and Russian. German seemed to be the third language.

Zagreb was started with bloodshed, a familiar story. The Croats lived on one side of the small stream, and the Zagrebs on the other. The street is still called Bloody Bridge Street. The cathedral, very tall, is Gothic. Smaller churches have "onion" tops as in Munich. Roofs are mostly of red tiles, I think. The main part of town has "old world" charm and is relatively tidy. Mike took me to the top of the tallest building for a panorama of the city. Later, I went in a largish book store and was able to get a small English-Yugoslavian dictionary with little trouble except the clerk short-changed me the equivalent of $1.20, probably by mistake, and she reimbursed me that amount.

My work at the Arkansas lab would be more closely related to Mike's than my West Virginia work. However, there was one close connection with my previous work at Leetown. I had worked on a *Tetrahymena* (ciliated protozoan) parasite, and that part of Europe has a similar one that kills carp fry.

Mike, his thirteen year old son and Jasna, Mike's lady veterinarian co-worker, took me to my first carp farm visit where they took us in a boat to ponds containing various ages of carp. They caught some with a throw net, including some with a visible virus disease. Then they served us a fine sausage-type lunch.

After that Mike took me to Plitvicka Lake National Park, about sixty miles south of Zagreb. The formation of the lakes may be unique. According to Mike, the microflora deposit calcareous clumps which build up on the natural dams, making them higher, the reverse of erosion! The waterfalls and lakes were some of the most beautiful I've seen. I hope my photos of them did justice! The carp farm was justifiable business, and we saw wild trout in the park so it was also justified as fisheries business. In the park we purchased large flat raised doughnut-type pastry which was good.

The countryside, except for a few modern cars and tractors, reminded me of France and Belgium during World War II. There were work horses, work oxen, small fields and much hand labor. I wished they could keep it that way.

In Yugoslavia I saw Trees of Heaven (a.k.a. Heaven Trees), *Ailanthus altissima*, which has invaded the U.S. In China, the leaves of Heaven Trees were used for silkworm feed, and apparently someone hoped to use them for the same purpose in U.S. However, in West Virginia and elsewhere, they became a nuisance because they are hard to kill and they crowded out more desirable native vegetation. At Mike's institute I gave a long seminar on American fish parasites and had a good visit with Mike's colleagues. For lunch that day they took me to an out-of-the way restaurant that roasted lamb and duck on a spit which was very good. Dinner was a fine home cooked meal at Fijan's. Merriam Fijan is a good cook.

The second side trip was to Scotland to visit Professor Ron Roberts at the Fisheries Institute at Stirling University and to give a seminar on American fish parasites. After I arrived at Edinburgh Airport, Ron's technician had me paged, the system for meeting passengers whose appearance was not familiar. Ron and family entertained me in regal fashion at their home near the institute. Afterwards Ron said "now my students can say they have met Dr. Hoffman as well as his boss, Dr. S. F. Snieszko." Ron's assistant gave me a review of world-wide fish farming. That afternoon I looked at Ron's student's unknown parasite

Figure 63. Howietown Fish Hatchery in 1974. Photo by the author.

which was still unknown that day! Later I gave a two hour slide show talk on fish parasites. I was urged three times to return for a sabbatical of a year or less, but it never materialized.

Stirling University of Scotland has the most beautiful campus I've seen. That opinion was probably influenced by a nice trout pond *on campus* where the pond water is cool enough year round for trout. Interestingly, I was able to understand all the Scots I talked with except one! Surprisingly, some of them had no noticeable Scottish accent.

Saturday morning we visited what may be one of the oldest fish (trout) hatcheries anywhere. The hatching house, resembling a small castle, was built in 1860 and must have been designed by a man with "vision" because it was still very functional even in 1974.

Next I flew to London where the hotel accommodations near Heathrow Airport were convenient and rather inexpensive. I had a "real" breakfast the next morning—bacon and eggs! The return to Little Rock, AR, was uneventful. Upon my return to the Arkansas lab I learned that my first Fish and Wildlife Service assistant, Dr. Robert Putz, was now a G.S. 15 (near the top) and was in charge of all USFWS research, a dramatic rise!

Soon after my return to the Arkansas lab, two lawyers from Michigan, representing one of the establishments involved in the accidental import of whirling diseased trout, flew in to obtain facts from me. To make a long story short, they won.

Europe 1975

This was a very memorable trip because wife Carolyn and son James, fifteen years old, accompanied me. On such a government trip it is permissible for relatives to accompany the conference participant as long as they pay their own expenses including travel, meals and lodging. The main event of the trip was for me to give a paper (talk) on protozoan parasites of fishes and to chair the fish disease module of the Wildlife Disease Association International Conference in Munich, Germany, in summer of 1975. The resulting chapters of the conference were published in a book, *Wildlife Diseases* (Proceedings of the third International Wildlife Disease Conference of 1975 (1976), Plenum Press, 686 pp).

This trip, mainly to Munich, Germany, was the most complete and complicated trip we have taken, before or after! The success of it was largely due to Carolyn's ability to plan and execute a complex itinerary. In short this is what we did: From Stuttgart, AR, we (Carolyn, Becky, James, I) drove to Iowa City, IA, to visit my major professor, Dr. L.O. Nolf, then to Strawberry Point, IA, to visit my parents and brother, then to Elmhurst, IL (suburb of Chicago), to visit Carolyn's sister and family. Here I digress a bit because, having been interested in rocks and fossils since my college geology course, we visited the Elmhurst Lapidary (rocks) Museum. Next we took Becky to O'Hare airport to fly to Washington, DC, to join her college roommate to go to Wesleyan University in Middletown, CT. Then Carolyn, James, and I drove to Guelph, Ontario, to meet members of the Wildlife Disease Society. Next we drove to Toronto and were guests of Dr. Alex Dektiarenk (a colleague and friend) and his wife before flying to Munich, Germany.

Because James was an organ student, we attended an organ concert in downtown Frauenkirche Cathedral, Munich, which was excellent. Surprisingly, a young man from our West Virginia home area recognized Carolyn, so we had a little down-home chat. The young man had been in Holland for the summer and was touring before going home. Small world indeed! A year or so later, when James was an exchange student in Bederkesa, Germany, he played the organ for services at a local church.

Figure 64. Dr. Schubert with the author outside Schubert's lab in Stuttgart, Germany, in 1978. Photo by Carolyn.

One side trip was to visit my counterpart, Prof. Dr. Gottfried Schubert (deceased) in Stuttgart, Germany. While there he showed me his lab and on-going research on aquarium fish parasites. Sometime later Dr. Schubert would serve one year as a consultant in the new large aquarium in Dubayi, UAE (United Arab Emirates).

A later kindness of Dr. Schubert occurred in September, 1977. A package was mailed from Wellington, New Zealand, addressed to Dr. G. L. Hoffman, Fish Farming Experimental Station, STUTT-GART, ARKANSAS, W. *Germany*. The package did go to Stuttgart, W. Germany, where the postal employees, noting "fish" in the address, phoned Dr. Gottfried Schubert, the research parasitologist at the University of Hohenheim, who accepted the package, addressed it correctly, and mailed it to me.

Later, Dr. Schubert and his daughter, who also spoke English, took us to a remarkable falconry at a castle near Heilbronn where a very large foreign vulture stole the show. We had two cars so Carolyn, James and I continued on north to Heidelburg Castle just for sightseeing and then back south to the German Black Forest, a very nice area with a castle, as usual. Then we returned to Munich.

We were scheduled to visit Prof. Dr. Jiri Lom, the world's expert on fish protozoa, in České Budějovice, Czechoslovakia, so on the way there we stopped in Vienna, Austria, to get our visas to Czechoslovakia. In 1975, before the Iron Curtain fell, this was not a quick task, so we stayed in Vienna overnight and, yes, had a chance to visit another castle. Next day we arrived at the Czech border on time, but were delayed a bit because I was arrested for possessing Czech money. We had read the directions wrongly and had some cash changed to Czech money before arriving in Czechoslovakia—bad! We were allowed to proceed, however, and I was later tried and found guilty in absentia and paid the penalty by mail!

We proceeded on to České Budějovice and arrived at the designated fountain on time, but no Dr. Lom! Carolyn, more proficient in foreign languages than I, finally found a Czechoslovakian who understood German, and she discovered that we were at the wrong fountain! We didn't know there were two fountains there. We hurried to the proper fountain and arrived just as Dr. Lom was preparing to leave because we were late! However, we had a nice visit including Dr. Lom's lab and the Veterinary Institute. We also visited another castle!

On the return to Munich we digressed to the jewel of a lake near Berchtesgaden, Germany. Berchtesgaden is a very pretty mountain town, but we saw no hint of Hitler's famous Eyrie.

We returned to Munich and then by air to Toronto. On the way back to Arkansas by car, we stopped at daughter Pam's place in Connecticut to celebrate our 27th wedding anniversary with her.

Figure 65. The author with Drs. Lom (right) and Cervinka (left) in 1975. Photo by Carolyn.

Europe 1978

Carolyn was able to accompany me to Europe again in 1978. The main purpose of this trip was for me to participate in the Fourth International Congress of Parasitology in Warsaw, Poland, 19-26 August, 1978. However our first stopover was in Bederkesa, near Hamburg, Germany, to visit son James' foreign exchange student "parents," the Hagenmullers, who had been very kind and helpful to James during his eleven-month stay in their home. We had rented a small Avis Fiesta Ford and continued south to Hannover to have lunch with James' other benefactor family, the Grassmuck's.

Then we flew to Stuttgart, Germany, to visit my colleague, Prof. Dr. Gottfried Schubert, again. We had been there in 1975. It was Sunday and my foot was hurting badly, so without luck we wasted the afternoon trying to find help. The next day Dr. Schubert took me to "his" hospital where diagnosis and treatment were made at no charge to me. Good friend! The dermatologist told me I had *Herpes simplex* (cold sore) on the bottom of one foot! The Betadine ("tamed iodine") helped greatly and my foot was soon much better. We had a quick tour of Gottfried's lab and the Aquarium and then drove a few miles north to have a quick look at the beautiful Rhine River and valley near Koblenz.

The Fourth International Congress of Parasitology was held in a huge "Stalinesque Gothic" Palace of Culture and Science which was very adequate. Many of my lifetime fish parasitology colleagues were there, including Prof. Dr. Oleg Bauer from Leningrad, Russia, Dr. Jiri Lom from Prague, Czechoslovakia, and Dr. Göran Malmberg from Stockholm, Sweden.

At the Congress, the multilingual Prof. Dr. Oleg Bauer (deceased) and I chaired a half-day session on fish parasites and had to prepare two written versions for the Congress. It was a pleasure to work with him. We had to give a 15 minute summary of our session in the Plenary session and had to have a long version ready for the Congress Program Chairman before leaving Warsaw. It was not easy, but we did it.

One morning at our hotel, each person was served one dill pickle and a cup of coffee for breakfast. Otherwise food was very good, especially the broiled duck one evening. However, about 80% of the 1500 participants suffered relatively mild travelers' disease, perhaps a stomach virus. Carolyn and I didn't catch it.

On our day off Carolyn and I took an all-day bus tour about one hundred miles north of Warsaw. We had a good look at their agriculture where most of the work (year 1978) was done by hand and horse, rarely with tractor. This way they could keep their small farms rather than become "collectivised" as they were in the USSR. When we stopped for lunch, a colleague and I noted a stream behind the restaurant. On our short walk there we came upon a fisherman who had a nice northern pike, *Esox lucius*, same species as in North America. However, that is probably the only native fish species existing on both continents. Some others, *Perca* and *Stizostedion*, are closely related to ours.

Carolyn always adapted well to foreign situations. In Warsaw, for example, she learned to manage the bus system quickly and went alone to Wilanew, a castle at the south end of Warsaw and up to the Old City at the north end. She found the latter charming because it had been restored just as it had been before the Nazis destroyed it. She learned that the whole city had been 80% leveled, the Old Town 90% and the ghetto 100%. They've done a tremendous job of rebuilding, and have

Figure 66. The Palace of Culture and Science, where the International Congress of Parasitology was held. Photo, dated 1978, by the author.

tried to make the ancient landmarks as accurate as possible. In one museum she noticed pictures of the Old Town that showed it undergoing restoration, and it looked almost exactly like the scene at the moment. It was dated about 1700 which tells the history of Warsaw in that area has been, apparently, bloody and sad.

She visited the lovely park in the southern part of the city where there is a striking monument to the famous composer, Chopin, which is a restoration. The originals of all the city's statues, she was told, were melted down by the Nazis for bullets.

One very nice thing for me was that I was finally able to meet several Polish fish parasitologists with whom I had corresponded for many years. Actually I had co-authored a fish parasite research paper with one of them.

Another side trip was to Stirling University in Stirling, Scotland, via London, England, and Aberdeen, Scotland. For this part of the trip we flew to Edinburgh, Scotland and rented a Chevette. We took the beautiful rugged heather-covered hill road to Aberdeen and returned via the shorter coastal road. At Stirling we were guests of Dr. Ron Roberts, Director of the Fisheries Unit and we had a nice visit with the staff of the Fisheries Unit. In Aberdeen we visited the National Fisheries and Unilever labs. It was always good to meet scientists that we had previously known only by correspondence and publications. As a boy in Iowa, I helped my dad fight bothersome thistles. In Scotland, the national flower is a thistle, a different species, of course, with a beautiful blossom.

The final side trip of 1978 was to Weymouth, on the southern coast of England, after flying from Edinburgh to London. We again rented a Chevette to drive to Weymouth. On the way we visited the famous spectacular Stonehenge and wondered how, without cranes or other heavy machinery, our ancestors could erect those huge monoliths. If I recall correctly it is believed that the stones served as a sort of clock, like a sundial. Also we passed by the beautiful Salisbury Cathedral. Interestingly, Salisbury is where my medical lab unit was supposed to go during WWII—our equipment did go there, but we, on another ship, went to Normandy, France. At Weymouth we visited the Weymouth National Fish Disease Lab and were guests of their histopathologist, Dave Bucke, whom we had met on a previous trip to England.

For our return to the London area, Dave Bucke had made reservations for us at the Harte and Garter hotel (a harte is a male deer) right across the street from Windsor Castle. We could look right over the wall and see a little of the inner compound. I had been inside the cas-

Figure 67. The author making a point during his presentation at České Budějovice in 1983. Photo by an unidentified colleague.

tle during WWII. We walked a little in Windsor Great Park. On our road we saw other estates with 5-foot stone or brick walls running for perhaps a mile along the edge of the park, and no sign of the mansion visible from the road, since it would be hidden by trees. By this time we had gotten used to driving on the "wrong" side of the road.

Czechoslovakia 1983

Dr. Jiri Lom's being probably the most famous fish protozoologist in the world, along with Dr. F. Moravec, nematodologist, being a leader in fish nematodes, made it appropriate for the First International Fish Parasite Conference to be held at their Institute in České Budějovice, Czechoslovakia, about 100 miles south of Prague, Czechoslovakia. It was wonderful to get together with the world leaders in fish parasitology, and particularly to chat with Dr. Oleg Bauer of Leningrad, Russia whom we hadn't seen since the Warsaw meeting in 1978.

For my presentation, I was given ten minutes more than originally assigned because Dr. Musselius of Moscow couldn't attend. For the first time I was also invited to speak at the closing ceremonies, a dis-

Figure 68. The author's initiation into the Czech Fish Culture Guild in 1983. Photo by Csabo.

tinct honor. Our colleague and friend, Dr. Gottfried Schubert of Stuttgart, Germany, had driven to the Conference and he kindly provided Carolyn and me with transportation. It was good to see him again, but he died suddenly two years later.

One evening, our host provided transportation to a nearby castle (Hluboka) perched on top of a low mountain. It was one of the choicest we visited, and houses a great collection of art works. In the art gallery we were treated to a piano duet concert, then later in the library we heard a delightful concert by a string trio in period costumes of the early 1800s, playing Czech music of that period.

The meals were a bit different than those in the U.S., but I recall one red letter dinner—a nice broiled trout. One night we had a carp roast picnic on the shore of one of the carp rearing ponds, as big as some lakes and artificially built in the 1300s in a remarkable feat of engineering. Dr. John Mackiewicz of University of New York at Albany and I were the Americans initiated into the Czech Fish Culture Guild by bending over a tub of carp and threatened with a rubber hatchet with the hatchet man reciting the ritual in Czech, of course! I felt that this was a distinct honor. The carp were roasted over an open fire of coals and were very good.

On the last free day, Dr. Lom led the entire group by buses around southeastern Bohemia with a visit to the lovely pink-stucco "water castle" of Cervena Lhota, less elaborate, but more livable than most castles. We were thrilled to see a stork's nest atop a tall chimney, and to watch the huge parent bird fly home to feed its nestlings.

I felt that this Conference was the beginning of a new era, providing better communication among the world's fish parasitologists. Unfortunately I retired two years later and did not witness the continuity of this wonderful project.

Australia and New Zealand 1986

I had retired in 1985, but was asked to chair a session at the International Congress of Parasitology at the University in Brisbane, Australia, so, with no government travel funds, Carolyn and I decided to attend and make it a science/vacation trip, our longest trip ever. We flew to Los Angeles, CA, and then overnight to Brisbane, Australia. It was our longest non-stop flight. The stewardesses were very kind, making us take off our shoes and occasionally providing hot, wet, face cloths for comfort. Brisbane is in northeast coastal Australia, but we landed at Sydney, on the southeast coast, for overnight hotel, then a charter bus for our parasitology group to Brisbane. This was good because the driver explained all the wonderful places as we passed by. The University at Brisbane is a large one, so we had excellent lecture rooms and accommodations. While I was at sessions, Carolyn explored the area, as usual, and enjoyed it very much. We were able to visit a nice natural park and a sheep ranch where, among other things, they demonstrated sheep-shearing for the wool. And, of course, we saw kangaroos and koala bears, the latter friendly and cuddly. The sheep-herding dogs were wonderful.

An Australian colleague, Dr. R.J.G. Lester, formerly from Canada, kindly invited Carolyn and me as well as another party to dinner at his home. His wife served a curry dish, well-liked by most people, but my system objected. However, our hosts were understanding and served me with a different entrée.

Figure 69. The author (overdressed) with colleagues on Heron Island in 1983. Photo by Carolyn.

After the sessions we took a week-long trip to Heron Island, a nice national park. We were taken there by helicopter, our only trip by such. It was a wonderful week; we had a little cabin and the meals were served delightfully in an open-air pavilion. I remember eating with colleagues from London, with "wild" birds flitting by. Our group was provided with canvas shoes for wading so we could see some of the interesting sea life up close. Also at Heron Island (incidentally, a heron is a stork-like fish-eating bird) we took a ride in the fish-viewing boat (with underwater glass windows) so we could see the native fish.

Next we flew to Auckland, in northern New Zealand, then traveled by chauffered car to Rotura in central part of the north island of New Zealand to a native Maori area where the native customs were explained to us and we were served excellent food. Along the way we visited a cave by boat and saw the spectacular glow worms (female beetle larvae) hanging from the cave ceiling, a real treat for us biologists!

All too soon it was time to go home. On that part of the trip we changed planes in our state of Hawaii (no sightseeing), then flew to San Francisco and home to West Virginia.

Mexico 1986

I was asked to give a series of lectures on freshwater fish parasites by Dr. Fernando Jimenez, Laboratorio de Parasitologia, Facultad de Ciencas Biologicas, at the Universidad Autonoma de Nuevo Leon, San Nicolás de los Garza, Nueva Leon, Mexico (really Monterrey, I think). Carolyn and I drove all the way on this trip, staying overnight in San Antonio, TX, with a friend from Grand Forks (1950s) days. The drive from there to Monterrey, Mexico, was very educational for us, desert-like with lots of interesting cactus plants. In a little village north of Monterrey we met, as scheduled, three students from the university, one of whom spoke English. We had lost most of the Spanish we learned in a class in Stuttgart, AR, some years earlier. They led us to a motel on the north side of the city which was very adequate with an excellent restaurant.

Next day we met our host, Dr. Jimenez, and we began to learn some of the habits of our new friends. If you are scheduled to meet at 8:00, it may be 8:30 or 9:00 before the meeting occurs. Lectures start when you get ready; the students saunter in and wait. Meals are any time; the evening meal may be at 8 or 9 pm. We had been advised to roll with the punches, and we did. We felt pretty well rounded by the end of the ten days there, but we enjoyed our kind hosts.

I was almost overwhelmed by the graciousness of my host and the students. Apparently they have great need of foreign literature and communication and wanted to thank me for coming to help. It was a little difficult for me to get used to a set schedule which immediately became overwhelmingly informal and uncertain. Other than that the professors and students were much like Americans. The facilities were adequate and they even had loud speakers to magnify my soft voice. I soon learned that I must allow twice as much time because half of the students did not know English well enough and Fernando had to interpret all that I said. Fortunately I had plenty of slides to help us over the language barrier. Discussion was good and I felt that overall the students were well prepared. Also the older fisheries employees in the second series seemed to be well prepared and dedicated to their work.

After my last lecture, two female students came up to thank me. One couldn't speak English; she asked the other to ask me if she could kiss me (for thanks) on the cheek. That was the only thanks of that kind that I received in my entire career!

Figure 70. The author with students at UANL in 1986. Photo by Carolyn.

I was able to dedicate both series at Monterrey to James. They apparently understood and Fernando added that part to the written reports about the series. I don't know where James got the feel of the need of our help in Latin America, but I sensed that it is really there. I don't know if I provided what they really had in mind, but I think so because of the courtesies they gave to me, including a standing ovation and the above-mentioned kiss. I was asked to sign all the diplomas and presented them after speeches by three dignitaries from the university and the Fisheries Department. After that they presented me with one of the diplomas with all of their signatures on the back.

The university is Universidad Autonoma Nueva Leon—the autonomous university of the state of Nueva Leon. Fernando explained the Autonoma part—they are politically entirely separate from the government. If we understood him correctly, their budget (in 1986) was greater than the entire city of Monterrey which is the third largest city in Mexico. I forgot to ask him who is the boss of the college president. The dean of biological sciences was a whopping 33 years old, but I think that is rather unusual there as well as here.

I guess the thing that struck me most about the people was their outward, and I believe, sincere, friendliness. I ran across only one disgruntled professor—he was having trouble getting funded properly. Somehow my rock hounding became known, and Fernando insisted that I take his entire collection of small colored stones. However I did not express a desire for his colorful soccer jersey with Ciencias Biologicas UANL (Biological Scientists of UANL) on it, but he insisted that I take it—it was one he had played soccer in. Carolyn washed it but I don't know what to do with it; Fernando is larger than I am.

The Escort behaved perfectly; it used no oil and got 32 miles per gallon. The entire trip was 4600 miles and we would probably not drive if I had to go back. The expense money they gave me came close to covering the cost for both of us, but not mileage on the car. The food at the nice Mexican motel was excellent with a choice of "hot" or not food. Beef was prominent on the menu, but I guess the food highlight for us was the excellent fruit-cocktails and salad. The papaya was especially good, I thought, but also sweet oranges, pineapple and melon. It was an educational and good trip.

Vancouver, British Columbia, Canada 1988

I was invited to give a progress report on whirling disease (*Myxobolus cerebralis*) of trout in the USA at the meeting of the International Fish Health Conference of the Fish Health Section of the American Fisheries Society in Vancouver, B.C., Canada in July, 1988. This disease was discovered in Europe in 1893 and has since been spread around the world with shipments of frozen cultured and wild trout. Diagnosis is usually straightforward but can be complicated by the presence of other *Myxobolus* species. An intermediate host, *Tubifex tubifex*, was discovered by K. Wolf, M. Markiw and J.K. Hiltunen (1986, *J. Fish Dis.* 9:83-85) which aids greatly in the control in fish hatcheries.

Our host in Vancouver was Dr. Hilda Ching who was researching fish tapeworms. She, husband and son were very gracious hosts. Hilda took us to many sights in Vancouver, too numerous to cite here.

While in Vancouver, on our day off, we took a side trip to nearby Nanaimo to re-visit the Pacific Biological Station by taking the ferry to Vancouver Island. My main contacts (friends, colleagues) there were Dr. Leo Margolis (deceased) and Dr. Z (Bob) Kabata. Leo was the Director and Bob the probable leading fish copepod expert of the world. After the conference, Carolyn and I rented a small car so that we could visit the Canadian Rocky Mountains and fly home from Calgary, Alberta, Canada. This was our first and only trip to the Canadian Rockies. Our route to Calgary took us through Kamloops, Revelstoke, Golden, Banff National Park to Jasper, then back south to Lake Louise with an overnight stay in Banff. Because we loved mountains, mountain streams and wildlife, we considered this our most wonderful trip. Also, it was the first time we were close enough to a glacier to walk up to it, touch it and see the effect of global warming!

Travel Can Be Broadening

Before WWII I had traveled from Iowa only to northern Minnesota and Moline, IL, via Davenport, IA. During WWII I served in, or near, Neosho, MO; Oxford, Cambridge, Plymouth, Salisbury, and the Biological Station at Lake Windermere, England; Paris, France; Munich and Darmstadt, Germany; and Antwerp and Liege, Belgium.

During my career, 1950-1985, my work took me to San Juan and El Yunque, Puerto Rico; Toronto and Vancouver, Canada; Monterrey, Mexico; London and Weymouth, England; Aberdeen and Stirling, Scotland; Paris, France; Copenhagen, Denmark; Stockholm, Sweden; Munich and Stuttgart, Germany; Sydney and Brisbane, Australia; Prague and České Budějovice, Czechoslovakia; Zagreb, Yugoslavia; Torino, Italy; and Leningrad (now St. Petersburg), Russia. Along the way I passed through Austria, New Zealand, and Switzerland.

After retirement I was invited to be a visiting professor at the University of Cork, Ireland, but couldn't make the trip for health reasons. I think Carolyn and I would have enjoyed it very much.

Within the U.S. I gave reports or consulted in various places in Alabama, Arkansas, California, Colorado, District of Columbia (Washington), Georgia, Idaho, Iowa, Kansas, Kentucky, Louisiana, Maine, Maryland, Massachusetts, Michigan, Mississippi, Missouri, Nevada, North Dakota, Oklahoma, Tennessee, Texas, Utah, Vermont, Washington, West Virginia, and Wisconsin.

18. Domestic Vacation Trips, Some Coupled to Parasitology Meetings

Skowhegan, ME and Strawberry Point, IA

After we moved to Grand Forks, ND (the site of the University of North Dakota), our month-long summer vacations usually alternated between Iowa (my parents) and Skowhegan, ME (Carolyn's grandparents). Her grandparents, Elmer and Mina Harris, were always very kind to us, as was their daughter Freda, who had retired as Dean of Women from the University of Vermont at Burlington. Previously she had been Professor of Mathematics there. Also very nearby in Skowhegan were Freda's Aunt Myrty Lord and cousins Priscilla and Marian Lord. All Harris' and Lords' were very nice people. It was always a pleasure to visit them. Grandpa Harris had retired from managing the local creamery.

On our first trip to Maine we traveled in tandem with Dr. John Davison (chemist) and his family, stopping for meals and lodging together which was nice. Also in those days, the 1950s, there wasn't much traffic. The Davisons and we have remained good friends ever since. Also later when Dr. Davison had transferred to the University of Virginia Medical School, they befriended our daughter Pamela when she was a medical student there. They, also, have remained good friends.

Before our children were born, I spent much time fishing in the Kennebec River, which ran through Skowhegan, and in the Sandy River, about 15 miles west. The Sandy was excellent for smallmouth bass. I first learned about the Sandy when, about 1951, I asked a local person about it; I was told "No, it's no good because it's full of bass since they stocked it with bass some years ago." So, from that time until Carolyn died in 1993, that was my favorite fishing stream in spite of the fact that I preferred trout. I always fished with a light fly rod and I recall four memorable occasions. Once, in the Sandy, where the bass were usually eleven to twelve inches long, I caught a three pounder. Once Grandpa Harris suggested I try a certain lake, about six miles east where there were nice pickerel. He was right; I caught several nice ones. On one camping trip in Maine, we stopped for lunch near the headwaters of the Sandy River. It was a rather small stream there so

Figure 71. The author with a nice catch of Sandy River smallmouth bass in 1977. Photo by Carolyn.

I walked up it about a quarter of a mile and found a nice pool with a nice brown trout in it which I caught on a small spinner fly. It was probably about fourteen inches long and made a nice dinner for us.

On one trip from Grand Forks to Maine, about 1951, going across southern Ontario, we stopped at Lake Nipissing for lunch. While walking near the beach we noted an animal burrow, rabbit or woodchuck size, and inside was a full grown brown weasel, staring at us, in curiosity I presume. Weasels are very cunning animals, and of course it reminded me of my boyhood trapping days. Also, Carolyn, a zoologist, was always interested in the things of nature.

Figure 72. My landlocked salmon. Photo by Carolyn.

Years later when our four children were large enough for camping we were camping near Moosehead Lake in northern Maine and I spent a half hour or so flyfishing for landlocked salmon in one of the feeder streams to the lake. I was lucky to hook and land one on a very small fly. It made a wonderful supper for us that night. Before the children were born, Carolyn enjoyed fishing, but that ended when they were born. They kept her very busy. All of those Maine relatives have died except at last account, the nephew of the cousins mentioned was living in Waterville, ME. However, I believe that many of Grandmother Mina Wyman Harris' relatives still exist.

Figure 73. The cottage on North Pond in 1999. It looked the same throughout the half-century we visited, except for the outdoor shower behind the enclosure on the left side. Photo by the author.

In addition to the wonderful hospitality of Elmer and Mina Harris, we were always welcome to stay at their cottage on North Pond, Smithfield, ME, nearby. Carolyn and I had 2 good aluminum canoes and she always enjoyed paddling to the small island about a quarter of a mile from the cottage. Also the cottage had a very nice sandy beach. On our last trip to Maine in 1992, it was Carolyn's wish to once again canoe to the island. She died the next year.

On our Iowa vacation trips, we were able to enjoy staying at the home of my parents, Ernie and Viola, in Strawberry Point. After my parents died, we enjoyed stays with my brother Melvin, his wife Eileen and daughter Linda at their home, also in Strawberry Point. Hikes in Backbone State Park were a staple of these visits.

Glacier National Park, 1956

The year before we moved from North Dakota to West Virginia, when Pam was six, Lyle five, and Becky five months, our vacation took us to Glacier Park. I've already mentioned the fine cutthroat trout I caught there. The older children enjoyed the lakes and mountains. We decided to climb a small mountain and Carolyn always remembered that she carried baby Becky, and I have always remembered that I carried her. It may be that we took turns and we both remembered how heavy she seemed!

We stopped at one nice lake on a nice day where Pam and Lyle put on swim suits and played in the water. The beach was rocky, not sandy, but the rocks, about four to five inches in diameter, were made very smooth by wave action. They were red-brown in color; I later tested them in the lab and found they were limestone, colored with iron. We have a nice photo of that spot, and also a nice one of an elegant waterfall.

Tucson, AZ, via Yellowstone Park and Grand Canyon, 1967

This was one of our greatest trips because the children were now 7, 10, 15, and 16 years old. The destination was Tucson, AZ, where I attended the annual meeting of the American Society of Parasitologists and where I could announce the first edition of *The Parasites of North American Freshwater Fishes*. Our 1964 Ford Fairlane station wagon was in good shape and we pulled a pop-up tent trailer. Carolyn, as usual, did well with the camp-stove meals and Lyle (15) and I managed the tent details. The others also had assigned chores.

I recall flyfishing in Yellowstone Park. Son Lyle caught a nice cutthroat trout and a fisherman from SE United States who preferred crappie to eat, gave us several trout for dinner. They were excellent.

I still have projection slides of the entire trip including Badlands of SD, Mt. Rushmore, Yellowstone Park, Dinosaur Monument, Bryce Canyon, Grand Canyon, Cliff Dwellers, Montezuma Castle, Desert Museum near Tucson, Painted Desert, Petrified Forest, Bryce Canyon, Flaming Gorge and CACTI—notably the huge Saguaro, but also Barrel, Beaver Tail, Organ Pipe, Pin Cushion, Staghorn, and Yellow Prickly Pear. I even have a slide of a roadrunner (bird) in Arizona. I've always been very grateful at how well the family handled the rigors of travel camping, particularly Carolyn who did so much of the work.

Figure 74. Dr. Landolt in her University of Washington office in 1969. Photo by the author.

Seattle, WA, via Glacier National Park and Other Points, 1969

I was to give a workshop on whirling disease, *Myxobolus cerebralis*, of trout at the University of Washington, so we visited Grand Forks, ND, Glacier National Park, Grand Coulee Dam, the Wenatchee Valley of Washington, and Olympia National Forest on the way with our 1964 Ford Fairlane station wagon and tent trailer. In Seattle we went up the Space Needle and visited a colleague, Marsha Landolt (deceased), Dean of the Graduate School, University of Washington. As mentioned above, she had been a histopathologist at our Leetown Lab, WV. At my workshop on the cause of whirling disease of trout, I was pleased to have my son Lyle, a high school student, as a participant.

Another highlight of the Seattle trip was a re-visit to Glacier National Park, one of the best in the world, I believe. At one point while Carolyn and the girls were hiking, Lyle and I walked up a small tributary till we found a nice pool and caught some large rainbow trout for supper. Lyle was a good fisherman. We also saw some mountain goats on a far mountain crag.

One rather unpleasant event was a summer snow flurry while we were pitching the tent in Glacier Park. However, we were prepared so we didn't suffer much.

Wilderness Canoe Trip, Boundary Waters, Minnesota 1973

Among so many other things, this trip can be dedicated to Carolyn who obtained the canoe route information and made the reservation at the Superior National Forest, northern Minnesota, bordered on the north by the Quetico National Forest of Canada. Permits are required to keep the trails from becoming overcrowded. I don't remember where we got our nice Superior-Quetico Canoe Country map, but it was published by W.A. Fisher Co., Virginia, MN 55792.

Five of our six family members made the trip: Carolyn (47), Lyle (21), Becky (16), James (13), and I (51). Pamela had a summer job and couldn't join us. We needed to take our own canoes so we strapped both, a 16 footer and 13 footer, to the top of our 1964 Ford Fairlane station wagon. Our tent for the eight day canoe trip was a lightweight nylon one.

Figure 75. En route to the Boundary Waters in 1973. Photo by the author.

CHAPTER 18

The start of our canoe trip was on Moose River about twenty five miles north of Ely, MN. The Moose River was rather small and would take us to Nina-Moose Lake on the start of the six day float. After crossing the lake we re-entered the same river and continued to Lake Agnes. Then our route was circular, passing through the southern part of Lake La Croix to Pocket Lake, Gebeonequet Lake, Green Lake, Rocky Lake, Oyster Lake and back to Nina-Moose River, then through Nina-Moose Lake and south to where we had parked our car.

We had no rain on the entire float and we camped at bare necessity camp sites, maintained by the National Park Service. The "facilities" were wooden and open air. The National Park Service also had periodic surveillance from an airplane. There was a nice picnic table at the first stop, but none after that.

On a cliff overhanging the water on Lake La Croix were faded paintings (pictographs) done by native Americans many years ago. We presume they were done when the lake was frozen. At one point between two lakes we passed through a magnificent "sea" of water lilies before Pocket Lake.

There were only two rather short portages; one between Gebeonequet Lake and Green Lake and between Green Lake and Rocky Lake. It was then that I discovered that my two sons were stronger than I because they carried the canoes easily!

A great joy for me, as a father, was that each family member, except me, sort of gravitated into one of the major camp chores. Carolyn took care of dishes and food, Lyle prepared the camp fires, James put up the tent, and Becky prepared the sleeping areas. I was the photographer. There was only one minor sibling squabble which was between the two boys in the smaller lightweight canoe. When I announced that we would change canoes if the squabble continued, the squabble discontinued.

Fishing was good. James caught the first one, a nice walleye, but after that Lyle seemed to have better luck with nice northern pike. We had underestimated the protein part of our diet which we made up with northern pike, roasted over the fire. Becky's digestive system rebelled and, to this day, she dislikes fish to the extent of not allowing it in their family meals.

We didn't see any bears, but enjoyed seeing eagles, as well as otter fishing. For me, it was one of our best family outings, perhaps partly because we were the only ones we encountered on this wonderful circular trip.

19. Retirement 1985–[2010]

In 1985, at age 67, I retired from the U.S. Fish and Wildlife Service at the Stuttgart Lab, AR, and Carolyn retired from nursing at the Stuttgart Hospital where she was on the nursing staff. We had kept our home near the Leetown, WV, National Fish Disease Lab, and were anxious to return. Shortly before our return, Carolyn was diagnosed with lymphoma cancer at the M. D. Anderson Cancer Institute in Houston, TX, and treated with x-ray and chemotherapy at Little Rock, AR. Rather soon all signs of lymphoma were gone, and I assumed she was cured.

It was good to be back in our own home that Carolyn had designed, including the acre of flower and vegetable gardens. She tended the flowers and I raised good crops of vegetables, including strawberries and raspberries. It was fun. What we couldn't use I gave to the neighbors and the lab crew a quarter of a mile away.

During the early retirement days Carolyn started Lamaze childbirth classes (birth without anesthesia) in our house and later continued the same at nearby City Hospital in Martinsburg, WV. Later she established a breast feeding class there, stressing the desirability of saving mothers' milk before birth as well as breast feeding. This became well established, and after her death the breast feeding room was dedicated to her along with a nice photo of her.

In 1983 and 1986, we took combined parasitology meetings and vacations to Australia/New Zealand and Mexico (both discussed earlier). In 1986-87 we had a year-long German exchange student, Mirjam Ziegelhoefer (now Herbst), to whom I devote a separate section, below.

In 1990 Carolyn went with a church group on a two week trip to Austria, Germany and Italy. I stayed home to work on the revision of my *magnum opus, Parasites of North American Freshwater Fishes*. Highlights for Carolyn, as evidenced by her photo-slides, were especially Oberammergau, Austria, with the famous Passion Play, and visiting our German exchange student, Mirjam Ziegelhoefer and her parents in Munich, Germany. She also enjoyed Merano and Venice, Italy, two castles, Lichtenstein Castle and Neuschwanstein Castle, both in Germany, and many other sites. It was a good trip for her and she was dubbed "the translator" by her tour group because of her help with foreign languages.

Because I enjoy flyfishing, one great gift to me, a long time employee of the USFWS, was the freedom to fish in the USFWS trout pond reserved for senior citizens, the physically challenged, and children when it was open. It was becoming more difficult for me to wade in streams so this was a very nice gift to me. And I did catch some very nice fish. Also touching to me was the fact that this pond was started by my former assistant, Dr. Robert Putz, when he had a tour of duty there during the construction of the new lab. Prior to this Dr. Putz allowed me to fish in the pond for cast-off experimental trout at his Freshwater Institute supported by the Conservation Fund, near Shepherdstown, WV. I was also able to teach several people to flyfish at that site. It was a real treat for me.

Of much importance to us, particularly during this period, were our pets. While we were still in Arkansas, a friend of Carolyn's gave her what was supposed to be a beagle dog, but I suspect that the father was a larger hunting dog because our beagle was much larger than the usual rabbit-hunting beagle. He was markedly undisciplined in many ways, for example, often eating paper products and other trash so we named him "Trouble." Otherwise, he was a dear friend, and Carolyn and he enjoyed many walks together. He died suddenly of unknown cause when he was about six years old.

We also had two cats from the animal shelter. One was nearly all white, but with some orange. We couldn't think of a good name for her so she was known as "Kitty" for her 21 years. The other, with tiger-stripe markings, was named "Tiger." Both were very good pets. Tiger disappeared, probably killed by a fox in the nearby thicket. He was a dear friend. His replacement also has tiger stripes and is officially "Tiger II." He still keeps me company at this writing (2008). All of my 3 surviving offspring are also "cat people."

Ten years after returning to WV (1992), Carolyn's lymphoma returned with a vengeance. This time it was most vigorous in the stomach. Despite excellent help she died in 1994, at age 68, at the University of Virginia Hospital where our daughter, Pam, had received her medical degree!

During this period, in addition to our German Foreign Exchange student, we had a summer student from Spain, Javier Fernandez. He was a well-mannered young man who fitted in well in our home. We took him to our cottage on a lake in central Maine, stopping on the

Figure 76. Tiger II in a characteristic pose. Photo by the author.

way at our son's home in Easton, PA and our daughter's home in Bridgeport, CT.

The year before Carolyn's death in 1994, she suggested a visit to our cottage in Maine. We went in October and fortunately there was an unusual warm spell so she had a nice last visit to North Pond. She was particularly desirous of a canoe trip from our cottage to the nearby island. She had never been in a canoe before we were married in 1948, but she learned to love canoeing and especially enjoyed the Minnesota Wilderness canoe trip mentioned earlier.

Our German Exchange Student, Mirjam Ziegelhoefer 1986-87

Mirjam (pronounced Miryam), from Munich, Germany, lived with us for one year under the Youth for Understanding Program, a good one. She attended Jefferson High School, WV, for her senior year. In the beginning she knew which courses would fit in her German program, but she had to abide by our high school rules. She spoke English well and adapted to our system very well. Later she returned to Germany to finish the German high school requirements and prepare for nursing school, which she completed. She taught nursing afterwards.

Figure 77. Spanish exchange student Javier Fernandez with the author outside the cottage on North Pond in 1986. Photo by Carolyn.

Some years later she visited me in WV with her boyfriend, Christian Herbst, now her husband, while he was a candidate for a Masters in Chemical Engineering at Johns Hopkins University. They have three little girls and Christian works for a chemical company near Munich. Thanks to Youth for Understanding, we have a life-long friendship.

The Second Edition of Parasites of North American Freshwater Fishes, *1999*

Following the publication of the first edition in 1967, I continued to keep notes on file cards of all the North American fish parasites that I could. During the 1990s I began putting the revision together

Figure 78. Mirjam with Trouble at our home in 1987. Photo by the author.

in earnest, including illustrations, some in color. The text, about 2000 pages, was mostly in long hand. Carolyn typed the whole thing before she died. After corrections, Dr. Bob Putz's secretary, Barbara Rinehart, at the Freshwater Institute, retyped the entire manuscript. The color prints were assembled, and the entire package went off to Cornell University Press. The result was a rather substantial book, 8.5 x 11 inches, and 539 pages. The printing was well done, but Carolyn never got to see it. In year 2009 Cornell Press declared it a classic and planned not to let it go out of print.

CHAPTER 19

Medical Woes

During my retirement, a medical problem that started when I was in my fifties grew to haunt me. About age fifty I had a rectal infection. My hometown physician elected to treat it only with saline washes. Apparently he had read about the antibiotic-resistant bacteria and so chose not to use antibiotics. That may have been a mistake because the infection lasted a long time and probably was the cause of my bacterial prostatitis which persisted for many years and was probably the cause of my prostate cancer at age seventy-five. The cancer was classified as six on a scale of one to ten. My local urologist gave me two choices of treatment—x-ray therapy or radical surgery. I chose the surgery wherein the surgeon did remove all of the cancer, but left me 100% urologically incontinent, with bladder spasms and some malfunction of the anal sphincter.

At that time my wife, Carolyn, was being treated for recurring lymphoma cancer. She was a registered nurse and suggested that I decline my referral to a urologist who could implant an artificial urinary sphincter (AUS). But, after she died a year later, when I could develop no urological control, I decided to go to Johns Hopkins Hospital, upon referral to Dr. Jacek Mostwin, the AUS expert, by my brother-in-law, Dr. T.H. Wilson II, a retired surgeon who had been an associate professor of surgery at Johns Hopkins School of Medicine.

I found Dr. Mostwin to be a very kind and competent urologist. He indicated that I probably made a mistake in choosing radical prostatectomy instead of x-ray therapy. Not only was he very competent, he was also well-educated and helpful. At one point he mentioned being at Oxford University in England for two years, where he earned a D.Phil. in experimental urology to go along with his MD degree.

After eight years, the AUS stopped functioning, so I had to go back to Johns Hopkins Hospital where Dr. Mostwin found that the AUS had developed a leak in the reservoir balloon, a fault that Dr. Mostwin said had never before developed in his work (and he had inserted hundreds). I couldn't help repeating Murphy's Law: "If anything can go wrong, it will!"

Dr. Mostwin removed the faulty one and installed a new one in so-called overnight surgery. The new one performed well for four years and, all of a sudden, was not working well, causing urethral irritation. My son Lyle, head of the Physics Department at nearby Lafayette College in Easton, kindly took me, then age 87, back to Dr. Mostwin who could find nothing wrong with the AUS or my "plumbing." However, he taught me how to manage the slight change in the pump, a plastic bulb which, when pumped, pushes fluid from the cuff, the tube around the urethra, up into the reservoir allowing urine flow. It was not perfect, but he said I didn't need a new one. Again, I was pleased with such a thoughtful surgeon. At one point I called him my hero, and he said "I'm no hero!" Ironically, the inventor of the artificial urinary sphincter, now supplied by Pfizer Hospital Products, Minnetonka, MN, was killed by the crash of his own plane near Minneapolis about 1994.

One highlight during my last tenure at my "castle" in WV was being able to help a little at the nearby fish disease lab. The library staff, Vi Catrow and Lora Mckenzie, had been very helpful to me during the revision of my book on fish parasites.

About year 2000, Dr. Rocco Cipriano, bacteriologist at the Leetown lab, organized and chaired the second USA/Russian fish disease conference which was held at a hotel in nearby Shepherdstown, WV. Rocco kindly chose me to give the keynote address. Rocco was very good at organizing such conferences, and it was very kind of him to invite an octogenarian to participate.

After Carolyn's death and my unhappy prostate removal, I ended up in the hospital with rather severe depression. Medication helped, but I think my cats and the large garden eventually helped more. It was very gratifying to see my surplus vegetables being used by neighbors. At one time the strawberries were very abundant and delicious homegrown cantaloupe were always welcome. I was glad that some could go to the nearby lab where I had worked for seventeen years.

Summer Vacation Trips to Emerald Isle, NC

After son-in-law Todd Craig, PhD, and daughter Rebecca, BS, graduated from Wesleyan University, CT, they moved to Raleigh, NC, where Todd worked for Union Carbide and Rebecca for a pharmaceutical company for some years. They spent their vacations on Emerald Isle, NC, renting furnished housing on the beach. They enjoyed that so much that they've gone there for their vacation ever since. Through the many years they also invited Todd's and Becky's parents. My wife, Carolyn, wanted to accept their invitation, but she died without being able to attend. For many years after that, however, my daughter Pam and I did accept. Pam still does, but health problems prohibit me from attending.

Retirement Home, Kirkland Village, Bethlehem, PA, 2004–[2010]

At the times of my three rather major operations, my offspring, Pam in Bridgeport, CT, and Lyle in Easton, PA, had to drive many miles to help me, interrupting their busy schedules. Therefore in 2004 we agreed that I should move to Kirkland Village, a large retirement complex in Bethlehem, Pa, just eight miles from son Lyle's home in Easton, PA.

Here, I try to remain active. In summer I have a large garden and my surplus is well used by other residents here. I also gave two slide shows on my fish parasite career and also one on our 1971 Wilderness Canoe Trip. And, of course, writing this autobiography was no small task for one of 87–90 years!

I was able to continue flyfishing for rainbow and brown trout in nearby Monocasy Creek until in year 2006 my arthritic hip became worse. However, son Lyle took me to Evergreen Lake, a fee for fishing establishment, where I didn't have to walk very far and could flyfish for rainbow trout.

20. Epilogue

I had total right hip replacement in December, 2007, which was successful except for a minor pigeon-toed effect of that leg. Recovery was accompanied by physical therapy which helped. In January 2008, Dr. Frank Tamarkin of the St. Lukes Hospital complex in Bethlehem, PA, removed the faulty AUS and installed a suprapubic catheter, so I am on tubes the rest of my life. None of it is expected to be life threatening, however. The proof of the pudding is that I can still, at 89, go fishing!

APPENDICES

A. Lifetime Publications of Glenn L. Hoffman as Senior Author

Hoffman, G.L. 1949. Isolation of *Saprolegnia* and *Achlya* with penicillin-streptomycin and attempts to infect fish. Prog. Fish Cult. 11:171-174.

___. 1953a. Parasites of fish of Turtle River, North Dakota. Proc. N.D. Acad. Sci. 7:12-19.

___. 1953b. *Scaphanocephalus expansus* (Crepl.), a trematode of the osprey in North America. J. Parasitol. 39:568.

___. 1954a. The occurrence of *Ornithodiplostomum ptychocheilus* (Faust) (Trematoda: Strigeida) in fish and birds. J. Parasitol. 40:232-233.

___. 1954b. Polyvinyl alcohol-fixative-adhesive for small helminths and protozoa. Trans. Am. Microsc. Soc. 73:328-329.

___. 1955a. Notes on the life cycle of *Bunodera eucaliae* Miller (Trematoda: Allocreadiidae) of the stickleback, *Eucalia inconstans*. Proc. Iowa Acad. Sci. 62:638-639.

___. 1955b. *Neascus nolfi* n. sp. (Trematoda: Strigeida) from cyprinid minnows, with notes on the artificial digest recovery of helminths. Am. Midl. Nat. 53:198-204.

___. 1956a. The life cycle of *Crassiphiala bulboglossa* (Trematoda: Strigeida), development of the metacercaria and cyst, and effect on the fish hosts. J. Parasitol. 42:435-444.

___. 1956b. Unpublished research.

___. 1956c. Medical Parasitology Laboratory Mannal. Burgess Publ. Co. 98 pp.

___. 1957a. Studies on the life cycle of *Cryptocotyle concavum* from the common sucker and experimentally in the chick. Proc. N.D. Acad. Sci. 11:55-56.

___. 1957b. Unpublished research.

___. 1958a. Studies on the life-cycle of *Ornithodiplostomum ptychocheilus* (Faust), (Trematoda: Strigeoidea) and the "self-cure" of infected fish. J. Parasitol. 44:416-421.

___. 1958b. Experimental studies on the cercaria and metacercaria of a strigeid trematode, *Posthodiplostomum minimum*. Exp. Parasitol. 7:23-50.

___. 1958c. Experimental infection with strigeoid cercariae. J. Parasitol. 44:229.

___. 1959a. Studies on the life cycle of *Apatemon gracilis pellicidus* (Yamag.). Trans. Am. Fish. Soc. 88:96-99.

___. 1959b. Recommended treatment for fish parasitic diseases. Bur. Sports Fish. Wildl., Kearneysville, WV. Fish. Leafl. no. 486.

___. 1959c. Unpublished research.

___. 1960. Synopsis of Strigeoidea (Trematoda) of fishes and their life cycles. Bur. Sports Fish. Wildl., Fish. Bull. No. 175 60:439-469.

___. 1962. Unpublished research.

_____. 1963. Parasites of freshwater fish. I. Fungi. 1. Fungi (*Saprolegnia* and relatives) of fish and fish eggs. Bur. Sports Fish. Wildl., Fish. Leafl. no. 564, pp. 1-6.

_____. 1964. Unpublished research.

_____. 1965a. *Eimeria aurati* n. sp. (Protozoa: Eimeriidae) from goldfish (*Carassius auratus*) in North America. J. Protozool. 12:273-275.

_____. 1965b. Unpublished research.

_____. 1965c. The control of fish parasites. *In* Biology Problems in Water Pollution, 3rd Seminar, 1962, pp. 283-285.

_____. 1967a. Parasites of North American Freshwater Fishes. Univ. of California Press, Berkeley. 486 pp.

_____. 1967b. An unusual case of fish disease caused by *Ophryoglena* sp. Bull. Wildl. Dis. Assoc. 3:111-112.

_____. 1969. Fungi (*Saproglegnia* and relatives) of fish and fish eggs. U.S. Fish Wildl. Serv., Fish Dis. Leafl. no. 21. 6 pp.

_____. 1970. Intercontinental and transcontinental dissemination and transfaunation of fish parasites, with emphasis on whirling disease (*Myxosoma cerebralis*). Pp. 69-81 *in* S.F. Snieszko (ed.), A symposium on diseases of fishes and shellfishes. Am. Fish. Soc., Washington, D.C., Special Pub. no. 5. 526 pp.

_____. 1973a. Parasites of laboratory fishes. Pp. 645-768 *in* R.J. Flynn (ed.), Parasites of Laboratory Animals. Iowa State Univ. Press, Ames. 884 pp.

_____. 1973b. The effect of certain parasites on freshwater fishes. Verh. Int. Verein. Limnol. (Leningrad) 18:1622-1627.

_____. 1974. Disinfection of contaminated water by ultraviolet irradiation, with emphasis on whirling disease (*Myxosoma cerebralis*) and its effect on fish. Trans. Am. Fish. Soc. 103:541-550.

_____. 1975. Lesions due to internal helminths of freshwater fishes. Pp. 151-187 *in* W.E. Ribelin and G. Migaki (eds.), The Pathology of Fishes. Univ. of Wisconsin Press, Madison. 1004 pp.

_____. 1976a. The anchor parasite (*Lernaea elegans*) and related species. U.S. Fish Wildl. Serv., Fish Dis. Leafl. no. 46. 8 pp.

_____. 1976b. Whirling disease of trout. U.S. Fish Wildl. Serv., Fish Dis. Leafl. no. 47. 10 pp.

_____. 1976c. The Asian tapeworm, *Bothriocephalus gowkongensis*, in the United States and research needs in fish parasitology. Proc. 1976 Fish Farm. Congr., Annu. Conv. Catfish Farm Tex., Tex. A&M Univ., pp. 84-90.

_____. 1976d. Is *Ornithodiplostomum ptychocheilus* a pathogen? Fish Health Sect./Am. Fish. Soc. Newsl. 4:13.

_____. 1977a. Copepod parasites of freshwater fish: *Ergasilus*, *Achtheres*, and *Salmincola*. U.S. Fish Wildl. Serv., Fish Dis. Leafl. no. 48. 10 pp.

_____. 1977b. *Argulus*, a branchiuran parasite of freshwater fishes. U.S. Fish Wildl. Serv., Fish Dis. Leafl. no. 49. 9 pp.

___. 1978a. Ciliates of freshwater fishes. Chapter 8 *in* J.P. Kreier (ed.), Parasitic Protozoa, vol. II. Academic Press, New York. 730 pp.

___. 1978b. *Chloromyxum esocinum*: First time in North America. Fish Health Sec./Am. Fish. Soc. Newsl. 6:7.

___. 1978c. *Salmincola californiensis* marches eastward. Fish Health Sect./Am. Fish. Soc. Newsl. 6:7.

___. 1978d. Bodomonas concava, a cryptic cryptogam for crippling *Cryptobia branchialis*. Fish Health Sect./Am. Fish Soc. Newsl. 6:9.

___. 1979. *Henneguya*, spore concentration with plankton centrifuge and species identification. Fish Health Sect./Am. Fish. Soc. Newsl. 7:9-10.

___. 1980. Asian tapeworm, *Bothriocephalus acheilognathi* Yamaguti, 1934, in North America. (In English.) Fisch. Umwelt 8:69-75.

___. 1981. Recently imported parasites of baitfishes and relatives. *In* 3rd Annu. Proc. Catfish Farm. Am. Res. Workshop (Las Vegas), pp. 45-46.

___. 1984. Two fish pathogens, *Parvicapsula* sp. and *Mitraspora cyprini* (Myxosporea), new to North America. Symp. (In English.) Biol. Hung. 23:127-135.

___. 1985. Anchor parasite (*Lernaea cyprinacea*) control. Fish Health Sect./Am. Fish. Soc. Newsl. 13:4.

___. 1990. Whirling disease (*Myxobolus cerebralis*) a worldwide salmonid parasite. J. Aquat. Anim. Health 2:30-37.

Hoffman, G.L., H. Bishop, and C.E. Dunbar. 1960. Algal parasites in fish. Prog. Fish Cult. 22:180.

Hoffman, G.L., and C.E. Dunbar. 1961. Mortality of eastern brook trout caused by plerocercoids (Cestoda: Pseudophyllidea: Diphyllobothriidae) in the heart and viscera. J. Parasitol. 47:399-400.

___. 1963. Studies on *Neogogatea kentuckiensis* (Cable, 1935) n. comb. (Trematoda: Strigeoidea: Cyathocotylidae). J. Parasitol. 49:737-744.

Hoffman, G.L., C.E. Dunbar, and A. Bradford. 1969. Whirling disease of trouts caused by *Myxosoma cerebralis* in the United States. U.S. Fish Wildl. Serv., Bur. Sport Fish. Wildl., Spec. Sci. Rep., Fish., no. 427. 15 pp.

Hoffman, G.L., B. Fried, and J.E. Harvey. 1985. *Sanguinicola fontinalis* sp. nov. (Digenea: Sanguinicolidae): A blood parasite of brook trout, *Salvelinus fontinalis* (Mitchill), and longnose dace, *Rhinichthys cataractae* (Valenciennes). J. Fish. Dis. 8:529-538

Hoffman, G.L., and J.B. Hoyme. 1958. The experimental histopathology of the "tumor" on the brain of the stickleback caused by *Diplostomum baeri eucaliae* Hoffman and Hundley, 1957 (Trematoda: Strigeoidea). J. Parasitol. 44:374-378.

Hoffman, G.L., and J.B. Hundley. 1957. The life-cycle of *Diplostomum baeri eucaliae* n. subsp. (Trematoda: Strigeida). J. Parasitol. 43:613-637.

Hoffman, G.L., and J.A. Hutcheson. 1970. Unusual pathogenicity of a common metacercaria of fish. J. Wildl. Dis. 6:109.

Hoffman, G.L., S.L. Kazubski, A.J. Mitchell, and C.E. Smith. 1979. *Chilodonella hexasticha* (Kiernik, 1979) (Protozoa, Ciliata) from North American warmwater fish. J. Fish Dis. 2:153-157.

Hoffman, G.L., M. Landolt, J.E. Camper, D.W. Coats, J.L. Stookey, and J.D. Burek. 1975. A disease of freshwater fishes caused by *Tetrahymena corlissi* Thompson, 1955, and a key for identification of holotrich ciliates of freshwater fishes. J. Parasitol. 61:217-223.

Hoffman, G.L., and J. Lom. 1967. Observations on *Tripatiella bursiformis*, *Trichodina nigra* and a pathogenic trichodinid, *Trichodina fultoni*. Bull. Wildl. Dis. Assoc. 3:156-159.

Hoffman, G.L., and F.P. Meyer. 1974. Parasites of Freshwater Fishes: A Review of Their Control and Treatment. T.F.H. Publications, Neptune City, NJ. 224 pp.

Hoffman, G.L., and A.J. Mitchell. 1978. *Branchiomyces* rides again. Fish Health Sect./Am. Fish. Soc. Newsl. 6:10.

___. 1980a. *Branchiomyces* again! Fish Health Sect./Am. Fish. Soc. Newsl. 8:3.

___. 1980b. Unpublished research.

___. 1986. Some parasites and diseases of warmwater fishes. U.S. Dept. Interior. Fish and Wildlife Leaflet 6, 22 pp.

Hoffman, G.L., and B. Moore. 1981. Search for improved control for the anchor parasite (*Lernaea*). U.S. Dep. Inter., Fish Farm. Exp. Stn. Stuttgart, Annu. Rep., p. 10.

Hoffman, G.L., G.W. Prescott, and C.R. Thompson. 1965. *Chlorella* (Alga: Chlorophyta) parasitic in bluegills. Prog. Fish Cult. 27:175.

Hoffman, G.L., and R.E. Putz. 1963. Unpublished research.

___. 1964. Studies on *Gyrodactylus macrochiri* n. sp. (Trematoda: Monogenea) from *Lepomis macrochirus*. Proc. Helminthol. Soc. Wash. 31:76-82.

___. 1965. The black-spot (*Uvulifer ambloplitis*: Trematoda: Strigeoidea) of centrarchid fishes. Trans. Am. Fish. Soc. 94:143-151.

___. 1969. Host susceptibility and effects of aging, freezing, heat, and chemicals on spores of *Myxosoma cerebralis*. Prog. Fish Cult. 31:35-37.

Hoffman, G.L., R.E. Putz, and C.E. Dunbar. 1965. Studies on *Myxosoma cartilaginis* n. sp. (Protozoa: Myxosporidea) of centrarchid fish, and a synopsis of the *Myxosoma* of North American freshwater fishes. Protozoology 12:319-332.

Hoffman, G.L., E.V. Raikova, and W.G. Yoder. 1974. *Polypodium* sp. (Coelenterata) found in North American sturgeon. J. Parasitol. 60:548-550.

Hoffman, G.L., and G. Schubert. 1984. Some parasites of exotic fishes. Pp. 233-261 in W.R. Courtney and J.R. Stauffer (eds.), Distribution, Biology and Management of Exotic Fishes. Johns Hopkins Univ. Press, Baltimore, MD. 430 pp.

Hoffman, G.L. and C.J. Sindermann. 1962. Common Parasites of Fishes. Circular 144. Reprinted 1969. U.S. Dept. of Interior, Fish and Wildlife Service, 17 pp.

B. Lifetime Publications of Glenn L. Hoffman as Junior Author

Bauer, O.N. and G.L. Hoffman. 1976. Helminth range extension by translocation of fish. Pp. 163-178 in L.A. Page (ed.), Wildlife Diseases. Plenum Press, New York. 686 pp.

Davis, H.S., G.L. Hoffman, and E.W. Surber. 1961. Notes on *Sanguinicola davisi* (Trematoda: Sanguinicolidae) in the gills of trout. J. Parasitol. 47:512-514.

Giudici, John J., D. Leroy Gray, and J. Mayo Martin. 1980. Manual for bait fish culture in the south. Pp. 49. Hoffman's chapter. Pp. 34-45.

Healy, G.R., G.J. Jackson, J.R. Lichtenfels, G.L. Hoffman, and T.C. Cheng. 1976. Foodborne Parasites, Chapter 38 in Speck, M.L. (ed.) Compendium of Methods for the Microbiological Examination of Foods. Prep by APHA on Microbiological Methods for Foods.

Hendrickson, G.L. and G.L. Hoffman. 1979. Formation of the host cyst and associated pathology of *Ornithodiplostomum ptychocheilus* (Trematoda: Strigeoidea) in *Pimephales promelas*. Program and Abstracts, The American Society of Parasitologists, p. 68.

Ikezaki, F.M. and G.L. Hoffman. 1957. *Gyrodactylus eucaliae* n. sp. (Trematoda: Monogenea) from the brook stickleback, *Eucalia inconstans*. J. Parasitol. 43:451-455.

Jayasri, M., and G.L. Hoffman. 1982. Review of *Myxidium* (Protozoa: Myxosporea). Protozool. Abstr. 6:61-91.

Kent. M.L., and G.L. Hoffman. 1984. Two new species of Myxozoa, *Myxobolus inaequus* sp. n. and *Henneguya theca* sp. n., from the brain of a South American knife fish, *Eigemannia virescens* (V). J. Protozool. 31:91-93.

Lom, J., and G.L. Hoffman. 1964. Geographic distribution of some species of trichodinids (Ciliata: Peritricha) parasitic on fishes. J. Parasitol. 50:30-35.

___. 1971. Morphology of the spores of *Myxosoma cerebralis* (Hofer, 1903) and M. *cartilaginis* (Hoffman, Putz, and Dunbar, 1965). J. Parasitol. 57:1302-1308.

McCraren, J.P., M.L. Landolt and F.P. Meyer. 1975. Variation in response of channel catfish to *Henneguya* sp. infections (Protozoa: Myxosporidea). J. Wildl. Diseases. 11:2-7.

Meyer, F.P. and G.L. Hoffman. 1976. Parasites and Diseases of Warmwater Fishes. Resource Publ. 127, 1976. U.S. Dept. Interior, Fish and Wildl. Serv., 20 pp.

Mitchell, A.J. and G.L. Hoffman. 1980. Important tapeworms of North American freshwater fishes. U.S. Dept. Inter. Fish and Wildl. Serv., Fish Dis. Leafl. no. 59. 18 pp.

___. 1981. Preparation of live channel catfish for shipment to states requiring a health permit. Aquacult. Mag. 7:28-29.

___, C.E. Smith and G.L. Hoffman. 1982. Pathogenicity and histopathology of an unusually intense infection of white grubs (*Posthodiplostomum minimum*) in the fathead minnow (*Pimephales promelas*). J. Wildl. Dis. 18:51-57.

Morrison, C.M., G.L. Hoffman and V. Sprague. 1985. *Glugea pimephales*, Porter and Richardson, 1941, n. comb. (Microsporidia, Glugeidae) in the fathead minnow, *Pimephales promelas*. Can. J. Zool. 63:380-391.

Nagel, M.L. and G.L. Hoffman. 1977. A new host for *Pleistophora ovariae* (Microsporidia). J. Parasitol. 76:160-162.

Putz, R.E. and G.L. Hoffman. 1963. Two new *Gyrodactylus* species (Trematoda: Monogenea) from cyprinid fishes, with synopsis of those found in North American fishes. J. Parasitol. 49:559-566.

___, G.L. Hoffman, and C.E. Dunbar. 1965. Two new species of *Plistophora* (Microsporidia) from North American fish, with a synopsis of microsporidia of freshwater and euryhaline fishes. J. Protozool. 12:228-236.

___, G.L. Hoffman. 1966. Earliest susceptible age of rainbow trout to whirling disease. Progressive Fish Culturist. 28(2):82.

___, ___. 1966. *Urocleidus flieri* n. sp. (Trematoda: Monogenea) from the flier sunfish. Proc. Helm. Soc. Washington. 33(1):46-48.

Raikova, E.V., V.C. Suppes, and G.L. Hoffman. 1979. The parasitic coelenterate, *Polypodium hydriforme* Ussov, from the eggs of the American acipenseriform, *Polyodon spathula*. J. Parasitol. 65:804-810.

Sawyer, T.K., G.L. Hoffman, J.G. Hnath, and J.F. Conrad. 1975. Infection of salmonid fish gills by aquatic amebas (Amoebida: Thecamoebidae). Pp. 143-150 *in* W.E. Ribelin and G. Migaki (eds.), The Pathology of Fishes. Univ. of Wisconsin Press, Madison.

Schachte, J.H. Jr. and Glenn L. Hoffman. 1986. Mortality of muskellunge fingerlings attributed to a multiple tapeworm infection. Amer. Fish. Soc. Spec. Publ. 15:357-359.

Snieszko, S.F. and G.L. Hoffman. 1963. Control of fish diseases. Lab. Animal Care 13:197-206.

___, ___. 1971. Fish Diseases, pp. 97-99 *in* A Manual of Wildlife Conservation, R.D. Teague (ed.), The Wildlife Society, Washington, D.C.

The University of Nebraska-Lincoln does not discriminate
based on gender, age, disability, race, color,
religion, marital status, veteran's status,
national or ethnic origin,
or sexual orientation.

www.ingramcontent.com/pod-product-compliance
Lightning Source LLC
LaVergne TN
LVHW041626070426
835507LV00008B/463